WHY DID THE CHICKEN CROSS THE WORLD?

ATRIA BOOKS

A Division of Simon & Schuster, Inc.
1230 Avenue of the Americas
New York, NY 10020

First Atria Books hardcover edition December 2014

ATRIA BOOKS and colophon are trademarks of Simon & Schuster, Inc.

For information about special discounts for bulk purchases, please
contact Simon & Schuster Special Sales at 1-866-506-1949 or
business@simonandschuster.com.

The Simon & Schuster Speakers Bureau can bring authors to your live event. For
more information or to book an event contact the Simon & Schuster Speakers
Bureau at 1-866-248-3049 or visit our website at www.simonspeakers.com.

Interior design by Kyoko Watanabe
Jacket design by Laywan Kwan
Map by Duncan Walker/Getty Images; chicken and footprints by Shutterstock

Manufactured in the United States of America

10   9   8   7   6   5   4   3   2

Library of Congress Cataloging-in-Publication Data is available.

ISBN 978-1-4767-2989-3
ISBN 978-1-4767-2991-6 (ebook)

WHY DID THE CHICKEN CROSS THE WORLD?

THE EPIC SAGA OF THE BIRD THAT POWERS CIVILIZATION

Andrew Lawler

ATRIA BOOKS

NEW YORK   LONDON   TORONTO   SYDNEY   NEW DELHI

For Finian

But that free servitude still can pierce our hearts.
Our life is changed; their coming our beginning.

—Edwin Muir, "The Horses"

CONTENTS

INTRODUCTION

Follow the chicken and find the world.

    —Donna J. Haraway, *When Species Meet*

Add up the world's cats, dogs, pigs, and cows and there would still be more chickens. Toss in every rat on earth and the bird still dominates. The domestic fowl is the world's most ubiquitous bird and most common barnyard animal. More than 20 billion chickens live on our planet at any given moment, three for every human. The nearest avian competitor is the red-billed quelea, a little African finch numbering a mere 2 billion or so.

Only one country and one continent are fowl-free. Pope Francis I regularly dines on skinless breast bought in the markets of Rome since there is no room for a coop in the tiny state of Vatican City. In Antarctica, chickens are taboo. Grilled wings are a staple at the annual New Year's celebration at the South Pole's Amundsen-Scott Station, but the international treaty governing the southern continent forbids import of live or raw poultry to protect penguins from disease. Even so, most emperor penguin chicks have been exposed to common chicken viruses.

These exceptions prove the rule. From Siberia to the South Atlantic's South Sandwich Islands, the chicken is universal, and NASA has studied whether it could survive the trip to Mars. The bird that began in the thickets of South Asian jungles is now our single most important source of protein, and we are unlikely to leave the planet without it. As our cities and appetites grow, so does the population of, and our dependence on, the common fowl. "Both the jayhawk and the man eat chickens," wrote the American economist Henry George in 1879, "but the more jayhawks, the fewer chickens, while the more men, the more chickens."

Until recently I never thought to ask why this creature, out of fifteen thousand species of mammals and birds, emerged as our most important animal companion. My reporting took me to archaeology digs in the Middle East, Central Asia, and East Asia as I pursued the question of why and how our species abandoned the quiet hunter-gatherer life in favor of bustling cities, global empires, world wars, and social media. This mysterious and radical shift to urban life that began in the Middle East six millennia ago continues to transform the earth. Only in the past decade, for the first time in history, have more people lived in cities than the country.

When I heard that excavators working on an Arabian beach had evidence that Indian traders had mastered the monsoon to sail across the open ocean more than four thousand years ago, I pitched the story to a magazine. These adventurous Bronze Age sailors inaugurated international trade and helped spark the first global economy, carrying Himalayan timber and Afghan lapis lazuli to the great Mesopotamian cities as Egyptian masons put the finishing touches on the Giza pyramids. In my pitch, I mentioned to the editor that along with remains of ancient Indian trade goods, archaeologists had uncovered a chicken bone that might mark the bird's arrival in the West.

"That's interesting," the editor said. "Follow the bird. Where did it come from? Why do we eat so much of it? What *is* a chicken, anyway?" I agreed, reluctantly, and a few weeks later I arrived in a seaside Omani village as the Italian archaeology team working at the beach site was returning from an afternoon swim in the Arabian Sea. The chicken bone? "Oh," said the dig director, toweling his damp locks. "We think it was misidentified. It probably came from one of our workmen's lunches."

Since chickens didn't pull Babylonian war chariots or carry silks from China, archaeologists and historians have not given the bird much thought, and anthropologists prefer watching people hunt boar than feed fowl. Poultry scientists are fixated on converting grain to meat as efficiently as possible, not in tracing the bird's spread around the world. Even scientists who appreciate the importance of animals in the making of human societies tend to overlook the fowl. Jared Diamond, author of the bestseller *Guns, Germs, and Steel*, relegates the chicken to a category of "small domestic mammals and domestic birds and insects" that are useful but not worthy of the attention due, say, the ox.

Underdogs and unsung heroes are journalistic red meat. The chicken is so underestimated that it is legally invisible. Although its meat and eggs power our urban and industrial lives, it is not considered livestock—or even an animal—under American law if raised for food. "Chickens do not always enjoy an honorable position among city-bred people," E. B. White noted. If they thought of chickens at all, it was "as a comic prop straight out of vaudeville." Though Susan Orlean declared the chicken the "it" bird in a 2009 *New Yorker* article devoted to the popular backyard chicken movement, the dog and cat retain their joint title as most beloved pet.

If all canines and felines vanished tomorrow, along with the odd parakeet and gerbil, there would be much mourning but minimal impact on the global economy or international politics. A suddenly chickenless world, however, would spell immediate disaster. In 2012, as the cost of eggs shot up in Mexico City after millions of birds were culled due to disease, demonstrators took to the streets, rattling the new government. It was dubbed "The Great Egg Crisis," and no wonder, since Mexicans eat more eggs per capita than any other people. The same year in Cairo, high-priced poultry helped inspire Egypt's revolution as protestors rallied to the cry: "They are eating pigeon and chicken, but we eat beans every day!" When poultry prices tripled in Iran recently, the nation's police chief warned television producers not to broadcast images of people eating the popular meat to avoid inciting violence among those who could not afford grilled kebabs.

The chicken has, quietly but inexorably, become essential. Though it can barely fly, the fowl has become the world's most migratory bird through international imports and exports. The various parts of a single bird may end up at opposite ends of the globe. Chinese get the feet, Russians the legs, Spaniards the wings, Turks the intestines, Dutch soup makers the bones, and the breasts go to the United States and Britain. This globalized business extends to Kansan corn that plumps Brazilian birds, European antibiotics to stave off illness in American flocks, and Indian-made cages housing South African poultry.

"A commodity appears at first sight an extremely obvious, trivial thing," Karl Marx wrote. But analyze it and the commodity turns into "a very strange thing, abounding in metaphysical subtleties and theological niceties." As I pursued the chicken's trail around the world, I found it full of surprising metaphysical and theological implications. Emerging from the Asian jungle as a magical creature, it spread around the globe, performing as a celebrity in royal menageries, playing an important role as a guide to the future, and transforming into a holy messenger of light and resurrection. It entertained us as it fought to the death in the cockpit, served as an all-purpose medicine chest, and inspired warriors, lovers, and mothers. In traditions from Bali to Brooklyn, it still takes on our sins as it has done for millennia. No other animal has attracted so many legends, superstitions, and beliefs across so many societies and eras.

The chicken crossed the world because we took it with us, a journey that began thousands of years ago in Southeast Asia and required human help every step of the way. It slept in bamboo cages on dugout canoes moving down the wide Mekong River, squawked in carts pulled by oxen plodding to market towns in China, and jostled over Himalayan mountains in wicker baskets slung across the backs of traders. Sailors carried it across the Pacific, Indian, and Atlantic Oceans, and by the seventeenth century, chickens lived in nearly every corner of every settled continent. Along the way they sustained Polynesian colonists, urbanized African society, and staved off famine at the start of the Industrial Revolution.

Charles Darwin drew on it to cement his theory of evolution, and Louis Pasteur used it to create the first modern vaccine. Its egg, after more than twenty-five hundred years of study, remains the premier model organism of science, and is the vessel we use to manufacture our annual flu serum. The common fowl was the first domesticated animal to have its genome sequenced. Its bones ease our arthritis, the rooster's comb smooths the wrinkles on our faces, and transgenic chickens may soon synthesize a host of our medicines. Raising the bird also offers poor, rural women and their children vital calories and vitamins to keep malnutrition at bay, as well as an income that can help lift struggling families out of poverty.

The animal remains a feathered Swiss Army Knife, a multipurpose beast that provides us with what we want in a given time and place. This plasticity that makes it the most valuable of all domesticated animals has become useful in tracing our own history. The chicken is a kind of avian Zelig, and since it is an uncanny mirror of our changing human desires, goals, and intentions—a prestige object, a truth teller, a miraculous elixir, a tool of the devil, an exorcist, or the source of fabulous wealth—it is a marker for human exploration, expansion, entertainment, and beliefs. Archaeologists now use simple mesh screens to gather bird bones that can tell the story of how, when, and where humans lived, while complex algorithms and high-throughput computing make it possible for biologists to trace the chicken's genetic past, which is so closely tied to our own. And neuroscientists studying the long-abused chicken brain are uncovering unsettling signs of a deep intelligence as well as intriguing insight into our own behavior.

Today's living bird has largely disappeared from our urban lives, and the vast majority inhabits a shadowy archipelago of enormous poultry warehouses and slaughterhouses surrounded by fences and sealed off from the public. The modern chicken is both a technological triumph and a poster child for all that is sad and nightmarish about our industrial agriculture. The most engineered creature in history is also the world's most commonly mistreated animal. For better and worse, we have singled out the chicken as our meal ticket to the world's urban future while placing it mostly out of sight and mind.

The backyard chicken movement sweeping the United States and

Europe is a response to city lives far removed from the daily realities of life and death on a farm, and the bird provides a cheap and handy way for us to reconnect with our vanishing rural heritage. This trend may not improve the life or death of the billions of industrial chickens, but it may revive our memories of an ancient, rich, and complex relationship that makes the chicken our most important companion. We might begin to look at chickens and, seeing them, treat them differently.

Even as we grow more distant from yet increasingly dependent on the fowl, our ways of describing courage and cowardice, tenacity and selflessness, and other human traits and emotions remain firmly bound up with the bird. "Everything forgets," said the literary critic George Steiner. "But not a language." We are cocky or we chicken out, henpecked or walking on eggshells. We hatch an idea, get our hackles up, rule the roost, brood, and crow. We are, in more ways than we might like to admit, a lot more like the chicken than the hawk or the dove or the eagle. We are, like the barnyard fowl, gentle and violent, calm and agitated, graceful and awkward, aspiring to fly but still bound to the earth.

1.

Nature's Mr. Potato Head

Probably the Eskimo is the only branch of the human family which has been unable to profit from this domestic creature.

—William Beebe, *A Monograph of the Pheasants*

On a chilly dawn in a damp upland forest of northern Burma in 1911, thirty-four-year-old biologist William Beebe crouched in the soggy undergrowth as a village rooster crowed in the distance. In a clearing just beyond his hiding place, men and mules carrying rice and ammunition prepared to leave for the nearby border with China, which was then convulsed with famine and revolution. As the caravan moved off into the morning light and the thin tinkle of the harness bells faded away, wild pigs, vultures, doves, and local chickens entered the abandoned camp to scavenge for leftovers.

A few minutes later, a colorful bird with a sleek and slender body and long black spurs sauntered into the clearing. Peering through binoculars, Beebe watched transfixed as the rising sun pierced the woods and hit the bird's feathers. "Just for a moment he was agleam, the sun reflecting metallic red, green, and purple from his plumage," he writes. The domestic hens and roosters stopped to observe the stately newcomer pass. "They recognized him as something alien, perhaps as superior, certainly to be respected, for they took no liberties with him," Beebe adds. The wild bird feigned not to notice the other animals, pausing only to snatch a bite and eye a village hen, before vanishing with a regal strut into the woods.

Beebe followed, sliding his lanky body quietly across the wet ground. At the bottom of a gully he spotted the male bird in a clump of bamboo with a female, which clucked happily and scratched the soil for worms as the wild cock "allowed no fall of leaf or twig to escape him, and it was interesting to watch how, every second or two, he systematically swept the sky and the woods all about." Never, he notes, was the bird off its guard, and he seemed to possess an almost eerie extrasensory perception. A distant yowling cat brought them both to attention; then a squirrel stirred nearby and the pair quickly darted into the dense forest.

This experience left an indelible impression on Beebe, who would go on to become one of America's first celebrity scientists. The bird, a red jungle fowl, carried itself like "an untamable leopard; low-hung tail, slightly bent legs; head low, always intent, listening, watching; almost never motionless." Beebe, an adventurous ornithologist who had traveled from Mexico to Malaysia, was awed by this singular creature that is the ancestor of the modern chicken. "Once the real fowl of the deep jungle is seen," he writes, "it will not be forgotten."

If the chicken is so common that it is concealed in plain sight, the wild bird from which it springs is surprisingly mysterious. Few biologists have observed the red jungle fowl in its native habitat of South Asia, and most of our knowledge of it comes from studies conducted in zoos on specimens that look like the bird observed by Beebe but act more like their tame barnyard brethren. Since the chicken and red jungle fowl are the same species—both bear the Latin name *Gallus gallus*—they can breed with each other. The number of chickens that can mate with their sibling and ancestor soared with the increase in the human population from India to Vietnam in the succeeding decades, diluting the wild gene pool. Beebe's observations give us an invaluable glimpse of the wild bird that would become the chicken.

How this shy and sly creature transformed into the epitome of domesticity has long puzzled biologists. "Those birds which have been pointed out as the most probable ancestors of the Domestic Fowl, do not appear to be more tamable than the Partridge or the Golden Pheasant," notes a perplexed Edmund Saul Dixon, an English pastor who served as Darwin's poultry muse, in 1848.

Like all domesticated animals, the chicken began as a wild creature that gradually was drawn into the human orbit. The wolf that became the dog came to us in its search of scraps of discarded food, which we provided in exchange for protection. Wildcats fed on the mice that ate our grain stores in the ancient Near East, so both felines and humans tolerated one another. Pigs, sheep, goats, and cows began as our prey and eventually were corralled into herds. The chicken's story is more enigmatic. Did the fowl come to us, did we go to it, or did we, over time, grow used to each other's presence?

The word *domestication* comes from the Latin term meaning "belonging to the house," and it suggests that, like a servant or slave, a domesticated animal does our bidding in exchange for shelter, food, and protection. Biologists today, however, see domestication as a long-term and mutual relationship, with bonds that can never fully be dissolved. Even feral pigs, Australian dingoes, and the mustangs of the American West retain genetic traits inculcated over thousands of years of living with people.

Few animals bond with us. Out of twenty-five thousand species of fish, the goldfish and carp can be considered domesticates. A couple dozen of more than five thousand mammals are domesticated, and out of nearly ten thousand bird species, only about ten are at home in our households or barnyards. Elephants can be trained to carry logs, cheetahs taught to walk on a leash, and zebras harnessed to pull a carriage, but they are only temporarily tamed, reluctant visitors rather than full-fledged members of the extended human household. Individuals from these species must be tamed anew with each generation. The red jungle fowl, distrustful of humans and ill-suited for captivity, seems an unlikely candidate for launching our species' most important animal partnership. That is why Beebe's minute scrutiny of the wild bird in its native habitat is the starting place for charting the chicken's journey across oceans and continents.

His visit to Burma on the eve of World War I had nothing to do with chicken history, however. It was part of an urgent mission by conservationists to study and record pheasants that faced extermination thanks to women's hats and rubber tires. Hundreds of thousands of acres of prime pheasant habitat were then being cleared across South Asia to make way for vast rubber plantations to supply the burgeoning bicycle and auto industries. Meanwhile, the feathers of exotic birds were a popular fashion statement for hundreds of thousands of Americans and Europeans, and egrets, warblers, terns, and herons across the United States were decimated as a result. A small protest movement that began in Boston when two socialites met for tea and founded the National Audubon Society grew into a potent political force that led Congress to ban sales of native bird plumes.

The large millinery industry promptly turned to the jungles of South Asia, home to all but two of the world's forty-nine pheasant species, including the red jungle fowl. This family of birds has elaborate and brilliant plumage unmatched by other avian species. Bird lovers feared that entire pheasant species would vanish before they could even be cataloged. "Members of this most beautiful and remarkable group are rapidly becoming extinct," warned Henry Fairfield Osborn, the president of the New York Zoological Society. "The record of their habits and surroundings, which is important to the understanding of their structure and evolution, will soon be lost forever." Osborn and other worried New Yorkers turned to Beebe, the wunderkind of ornithology.

Beebe had dropped out of Columbia University to work at the recently opened New York Zoological Park in the Bronx, and he was only twenty-two years old when he designed its innovative flying cage. While other American zoos kept birds in small pens, this one was a breathtaking, open chamber, 150 feet long and 75 feet wide, soaring 50 feet into the air above a stream, plants, and trees. The flying cage became a central New York attraction after its 1900 inauguration. Rail-thin and with a dashing mustache, Beebe was adept at combining science with adventure, high society, and entertainment. He befriended Theodore Roosevelt, liked costume parties, flew World War I air missions, starred in documentaries, and descended three thousand feet into the ocean in a bathysphere. "Boredom is immoral," he once told a friend. "All a man has to do is see."

In 1902, Beebe married a wealthy and talented Virginia bird-watcher and novelist named Mary Blair Rice. With Osborn's encouragement and with financial backing from a New Jersey industrialist, they set out in 1909 from New York Harbor aboard the *Lusitania*, the ill-fated liner sunk six years later by U-boats that helped push the United States into the war against Germany. For seventeen months, the couple worked their way across the southern girdle of Asia, avoiding bubonic plague, fleeing a riot in China, and contending with bouts of Beebe's periodic depression. Their marriage did not survive the difficult trip. Upon their return home, Rice left for Reno and filed for divorce, accusing her husband of extreme cruelty. He went on to publish the four-volume *A Monograph of the Pheasants*.

The couple discovered that mass slaughter indeed threatened numerous species, given rubber plantations, the market for feathers, and Chinese adoption of a diet heavy in meat. "Everywhere they are trapped, snared, pierced with poisoned arrows from blowpipe or crossbow, or shot with repeating shotguns," Beebe wrote dispiritedly. He saw huge bales of silver pheasant feathers stacked in the customs house in Burma's capital of Rangoon and complained that Nepal and China exported large quantities to the West, despite new laws forbidding their import. The fast-expanding rubber plantations, he added, severely reduced habitats for the remaining birds.

Beebe was particularly taken with the red jungle fowl, "the most important wild bird living on the earth," given that it is the living source of all the world's chickens. He watched with astonishment as one fowl rocketed out of the brush to anchor safely on the high branch of a tree, while another soared across a half-mile-wide valley. "There is no hint of the weak muscles of the barnyard degenerate," Beebe states with a biologist's condescension toward domestic animals. Most of the red jungle fowl's days, however, are spent on the ground, feeding in the early morning and late evening and resting in the shade during the heat of the day. That rhythm was in synch with many early farming societies in the tropics.

Little was known of the bird's diet, so Beebe spent a good deal of time probing the digestive pouch near its throat—the crop—and picking through the guts. He found mostly remnants of plants and insects. Although an omnivore, the bird prefers grasses like bamboo shoots and live bugs to grain, herbs, or carrion. This would have made it, unlike crows or sparrows, a friend to early farmers.

Beebe also was struck by the sedentary and social nature of red jungle fowl, qualities that also likely appealed to ancient peoples. The birds rarely stray from their home turf, and mothers care for their chicks for nearly three months before they leave to form their own social groups. "It is seldom that I have seen or have heard of a solitary cock or hen," Beebe writes. Unlike other pheasants, jungle fowl prefer to roost together at night. The favored place to sleep is usually a bamboo stalk bent low. That might seem a poor choice, since it is closer to the ground than a tree branch and liable to sway in the wind, but few predators can climb the smooth stalks. An isolated tree is another favorite perch, less vulnerable to nighttime attack. While most birds chafe at being locked up at night, the red jungle fowl's sleeping habits and vulnerability lend themselves to a chicken coop.

The bird's predators are, after all, legion. Minks and jackals like the taste of wild chicken, as do hawks and eagles, while lizards and snakes enjoy the eggs. The fowl is not, however, a prolific egg producer like its domesticated sibling. Each hen lays an average of a half-dozen each year in carefully concealed ground nests, fewer eggs than many other pheasant species. Nor is the bird larger and fleshier than many of its cousins. The copious meat and eggs that mark the chicken today are solely a result of human intervention over millennia and not a characteristic of its ancestor. But the male red jungle fowl's ability to sense danger and crow a warning might have served as a handy alarm system for early human settlements.

There are three other species of jungle fowl—the gray, the green, and the Sri Lankan—and Beebe closely observed these as well. All share similar traits, but they live in a much more restricted geo-graphical area than the red, which thrives from five-thousand-foot mountainsides in the chilly Himalayan foothills of Kashmir to the steamy tropical marshes of Sumatra. From Pakistan to Burma to the Pacific coast of Vietnam, the red jungle fowl is at home in a remark-able variety of habitats, and has evolved into several distinct varieties specific to those climates. This capacity to adapt to a wide variety of climates and food gave the red jungle fowl the right stuff for a jour-ney that would take it to almost every conceivable environment on earth.

Beebe concludes that the red jungle fowl is made of a mysterious and unique kind of "organic potter's clay" that sets it apart from other birds, what he called "latent physical and mental possibilities." He was writing at the dawn of genetics, and the same year that he watched the wild cock strut across the Burma clearing, Thomas Hunt Morgan at Columbia University—Beebe's would-be alma mater—published a series of seminal papers in *Science* based on fruit fly studies that demonstrated the existence of chromosomes that carry specific genetic traits. The research helped launch the modern genetics rev-olution that Darwin had laid the foundation for a generation before.

The fowl's unusual plasticity, Beebe theorized, let humans mold it into the "beautiful, bizarre, or monstrous races" of the domesticated chicken. Plumage could be lengthened or shortened, colors and their patterns quickly altered, and the size of limbs extended or reduced. While the wild bird has a tail less than twelve inches in length, that of one Japanese breed stretches twenty feet. A domesticated rooster's comb alone can take more than two dozen distinct forms. Males could be altered to become fierce fighters with fewer feathers for an opponent to grasp. With tinkering, the two-pound red jungle fowl morphed into the twenty-ounce bantam and the brawny ten-pound Brahma, while a White Leghorn hen can churn out an egg a day.

The red jungle fowl, in other words, is nature's Mr. Potato Head. Its daily rhythms, diet, adaptability, and sedentary and social nature were the perfect match for humans. In 2004, a huge international team of scientists called the International Chicken Polymorphism

Map Consortium decoded and published the chicken genome, the first genetic map of a farm animal and potent proof of the bird's economic importance. The researchers discovered that the vast majority of the 2.8 million single-nucleotide polymorphisms—selected pieces of the genome that each represent a difference in a single DNA building block—likely originated before domestication. The modern chicken, in other words, is still mostly red jungle fowl; although that conclusion was based on the assumption that the red jungle fowl genes that were studied were in fact those of purely wild birds.

The results offered practical ways for breeding companies to create even larger and meatier birds through crossbreeding for particular genetic traits, but they provided frustratingly little insight into the changes that transformed the wild creature into a barnyard staple. Later research hinted that a mutation prompting fast growth might have put the red jungle fowl on its domestication track thousands of years ago, but there is little evidence that humans bred the bird, at least initially, primarily for food. What scientists need is a reliably pure red jungle fowl to tease out the minute differences that make one bird wild and one domesticated.

This is not as easy as it sounds. By World War I, exotic bird feathers on hats were out of fashion and the rubber boom had crashed. This gave the pheasants of South Asia, including jungle fowl, time to recover. During his expedition, however, Beebe noticed in passing that some male red jungle fowl lacked eclipse plumage, a set of purplish feathers that appear when a male sheds its red-and-yellow neck feathers and central tail plumage in late summer. In fall, the bird molts completely and grows a new set of feathers. Chickens skip the eclipse plumage phase, so Beebe saw this as a sign of "an infusion of the blood of native village birds" into the wild genome.

Nearly a century passed before another biologist realized that the ancestor of the world's most prolific bird and humanity's most important domesticated animal was slowly and inexorably vanishing, a victim of its own evolutionary success as Asia's expanding chicken flocks threatened to overwhelm the wild bird's genetic integrity. Its passing could blot out the first steps of the chicken's journey forever.

But thanks to an obscure U.S. government program designed to quiet the clamor of Southern hunters, the red jungle fowl may yet reveal its story.

Importing wild animals from distant and exotic lands is a practice as old as civilization. Early monarchs in the ancient Near East boasted of their menageries of lions and peacocks, a Baghdad caliph sent Charlemagne an elephant, and a fifteenth-century Chinese emperor showed off his giraffes to astonished diplomats. Since the vast majority of species are not as adaptable as chickens or humans to a new climate, diet, or geography, most transplanted animals quickly perish.

One of the few successful imports of a wild bird to the United States is China's common pheasant, also known as the ring-necked pheasant, which was brought from the Far East and proliferated in the Midwest and Rocky Mountains in the 1880s, though it steadfastly refuses to live south of the Mason-Dixon Line. Many other alien species that proliferated proved disastrous, such as European starlings and English sparrows, which eat crops, harass indigenous birds, and can bring down a jetliner. In the early 1900s, at the same time that Congress moved to protect native species from hat fashion, lawmakers banned import of potentially harmful species.

By the Great Depression, native wildlife of all sorts, from deer to ducks, was rapidly disappearing, and alarm spread among conservationists, hunters, and the gun and ammunition industry. In 1937, President Franklin Roosevelt signed bipartisan legislation providing the first regular funding for wildlife research designed to understand and address the problem. World War II put a halt to this work, and the emergency only deepened a decade later when millions of returning veterans took to the woods with high-powered rifles. Hunting seasons around the country were sharply curtailed and the entire Mississippi River flyway was set off-limits. "American wildlife management officials now are facing what is unquestionably the gravest crisis in the long and colorful history of wildlife conservation on this continent," warned the president of the International Association of

Game, Fish and Conservation Commissioners in an Atlantic City ballroom in 1948.

The chief of New York's game conservation department, a self-assured and newly minted PhD named Gardiner Bump, proposed a radical solution. A hulking man over six feet tall and weighing more than two hundred pounds, Bump argued that importing wild game birds from Europe and Asia to North America, if done scientifically, would replace the depleted stocks of native species. The director of the U.S. Fish and Wildlife Service was wary of introducing a potential pest, since he led an organization created largely as a result of the outcry against alien species. Desperate for ideas in the face of impending catastrophe, he reluctantly agreed.

Bump and his wife, Janet, set out on a two-decade-long search for the best candidates, traveling from Scandinavia to the Middle East. None of the dozens of game bird species they shipped to the United States adapted and proliferated on their own. Meanwhile, Bump's colleagues and superiors in Washington were under increasing pressure from Dixie lawmakers to find a bird to satisfy their disgruntled hunting constituency. Southerners had mainly duck and quail to hunt, and they were eager to bag more challenging game fowl like pheasant. In 1959, the Bumps rented a house in a well-to-do suburb of New Delhi with a backyard large enough to accommodate bird pens, betting that on the subcontinent they could locate a suitable Southern candidate.

Old British hands consulted by Bump urged him to focus on the red jungle fowl, which was secretive, smart, and fast and liked a warm, humid climate in a wooded environment. Bump assured Washington that he was on the trail of a promising species, but Indian civil servants denied his request to send an official expedition into the Himalayan foothills that were prime red jungle fowl habitat. In those days, India was friendly with the Soviet Union and wary of Americans close to its sensitive borders with Pakistan and China. Undeterred, Bump went on a private hunting holiday. Exploring the wooded hills and forests of northern India, where the Ganges River gushes out of the Himalayas, he was impressed by the challenge posed by the fowl. It was, he wrote, "almost as difficult to hit on the wing as the ruffed grouse." He decided to send out locals to net the birds and collect their eggs.

Bump had one overriding concern. He needed truly wild birds that would survive predators in the American South. If his imports were tainted with domesticated chicken genes, they might lack the shy and sly qualities of the fowl observed by Beebe, and therefore not last long enough to procreate. To avoid this problem, he directed that all the eggs and chicks of red jungle fowl had to be collected at least three miles from the nearest village. Later he maintained that most of the specimens were taken ten to fifteen miles from the closest human habitation, though verifying this claim a half century later is difficult.

The biologist died decades ago, but Glen Christensen, who worked with him in India as a young ornithologist, is still alive and pushing ninety. "Hold on, I have to get my oxygen," he says when I call him at his home in the Nevada desert. After a pause, he returns to confirm that Bump was well aware of the crossbreeding problem. Christensen laughs at my idea of a hardy and enterprising outdoorsman roaming the wild hills of the Hindu Kush with rifle and knapsack. "He wasn't too involved with the trapping. In fact, he wasn't much of a field man," he adds, taking another pause to inhale. "He sat in his compound in Delhi like an old country squire."

More difficult than trapping the birds was finding a way to get them from New Delhi to New York, a seventy-three-hundred-mile journey. Flights from India to the United States required a series of plane changes and took a total of four days, a logistical nightmare for anyone shipping wild birds. In 1959, Pan Am inaugurated Boeing's new 707 jet to reduce this time to one and a half days on the same aircraft. The Bumps held a lavish dinner party for Delhi airline agents, serving cocktails in the backyard among the sturdy pens while explaining their effort. Impressed, or possibly just drunk, the Pan Am agents agreed to help.

By May 1960, the Bumps were collecting red jungle fowl and their eggs brought by trappers. They hatched the eggs under domestic hens, placed them in backyard pens, and fed them a poultry mash commandeered from the American exhibit at the World Agricultural

Fair. Thanks to Pan Am, seventy were sent to four Southern states via New York. Later, in 1961, forty-five more were shipped to the United States. Meanwhile, state game managers bred the birds in special hatcheries, raising ten thousand red jungle fowl for release across the South, starting in the fall of 1963. The couple was hopeful that at last they had finally found a solution to the game fowl crisis.

The released birds, however, appeared to vanish in the Southern wild, victims of predators, weather, disease, or some deadly combination. Back in the States, Bump traveled peripatetically among state hatcheries for the rest of the decade, antagonizing game managers with his increasingly desperate demands. His critics, always legion in the conservation field, carped loudly that the effort to introduce foreign species was a waste of time and money. Wildlife populations had rebounded in the 1950s through a careful combination of hunting limits and habitat protection. The more insidious new threat, particularly to wild birds, was pollution. A former U.S. Fish and Wildlife Service employee named Rachel Carson, who was mentored by William Beebe, published *Silent Spring* in 1962. The bestseller propelled the environmental movement toward understanding and preventing the chemical pollution and habitat destruction that were taking a toll on native species.

In early 1970, as the nation celebrated its first Earth Day and President Richard Nixon prepared to organize the new Environmental Protection Agency, Bump picked up the phone in his Washington office and called a young biologist in South Carolina with a keen interest in red jungle fowl. The foreign game program was about to be canceled, and the remaining birds kept for breeding at state game facilities in the South would soon be destroyed. "They are going to assassinate the jungle fowl," he warned his junior colleague, I. Lehr Brisbin. "Save what you can."

Now in his midseventies, Brisbin lives with his third wife in a tony suburban neighborhood not far from the nuclear weapons laboratory where he has worked for half a century, just off a street lined with faux

Colonial houses and well-tended lawns. His driveway begins like the others, and then abruptly turns into an unpaved track descending into thick woods. A box turtle wearing a radio collar lumbers past as I ring the bell and Brisbin calls for me to come in.

He's sitting barefoot on the parquet floor of the foyer with a green knapsack and maps strewn around him. On the hall table behind him, a stuffed fox in a radio collar stares directly at me. "It just dropped dead?" he is saying into the phone. "Did you freeze him?" Pause. "Well, if your bird died it isn't going to bother me as long as you freeze him." He hangs up, grabs a wooden cane leaning by the door, and hoists his small, wiry frame upright. Brisbin has agreed to take me to see the descendants of the wild chickens that he rescued from destruction, birds that may prove to be the last of the world's truly wild red jungle fowl.

His first job as an ecologist in the late 1960s was to determine if chickens could survive the trip to Mars. To do this, he put a squawking fowl into a metal box and lowered it into a deep lead-lined pit containing a low-level radiation source at the government's Savannah River Site, where nuclear engineers made tritium and plutonium for weapons of mass destruction. Repeated exposure for a few minutes each day simulated the environment of outer space, beyond the protective blanket of the earth's atmosphere. The ninety birds he studied proved remarkably hardy even after a month of significant exposure to gamma radiation. None died. Growth rates slowed, but the skeleton remained largely unaffected except for a slightly shorter middle toe.

Poultry, he concluded, could survive the interplanetary trip. He published his findings in the same month that Neil Armstrong and Buzz Aldrin stepped onto the lunar surface. A bird accompanied those astronauts on their July 1969 mission to the moon, albeit in the form of freeze-dried cream of chicken soup. NASA managers dreamed of sending live animals with astronauts to settle the Red Planet, envisioning roosters crowing during a pink Martian dawn as self-sufficient pioneers established humanity's first extraterrestrial beachhead. Dogs and cats could wait, but chickens and their eggs were essential to the venture. The ecologist's research for the space agency was part of that grand plan, which never took off.

As a graduate student at the University of Georgia in Athens, Brisbin had studied the growth rate of chickens through their entire life cycle. The bird can live a decade or even two. Since those grown for meat or eggs are slaughtered at a young age, however, researchers knew little about the middle and elderly years. Brisbin realized that it would be useful to compare the chicken's life cycle to that of its wild ancestor, and he dreamed of going to India to see the wild chicken in its native habitat. Just as NASA never made it to Mars, Brisbin never got to the subcontinent. But a year after publishing his paper on outer-space chickens, he got Bump's anxious call.

Alerted to the birds' plight, Brisbin hopped into his Ford station wagon and drove two hundred miles to a Georgia game station, where he loaded up the car with a hundred red jungle fowl eggs. Two months later, he wrote Bump that he was raising thirty-five healthy young red jungle fowl in his chicken pens in the shadow of the nuclear facilities. He learned through trial and error that they were an unusually skittish lot, and he avoided touching the birds and limited their contact with people. A year later, despite his expertise and precautions, only eight birds were left. Two colleagues from the University of Georgia gave him sixty-nine additional red jungle fowl from the state game farm in Alabama, from the same stock that Bump had brought from India. That infusion helped stabilize the flock.

In 1972, Brisbin was transferred to a desk job in Washington. He could not take the fowl to the capital, but he also couldn't find anyone willing to take care of the temperamental birds. The Bumps had retired to their upstate New York farm, ecology colleagues at Savannah River scoffed at Brisbin's hobby, and nuclear engineers were embarrassed by the presence of low-tech fowl on their high-tech campus. Then, "out of the blue, Isaac Richardson called," says Brisbin. An eccentric loner and wealthy owner of a beef-and-pork slaughterhouse in Tuscaloosa, Alabama, Richardson sold meat for profit but raised exotic birds for pleasure.

Having heard of Brisbin's plight, Richardson went to Savannah

River that June, took a dozen birds home, and reported back that they were thriving. Encouraged, Brisbin put the rest of his birds in a shallow box padded with foam so they wouldn't bash themselves to death and drove them to Alabama. It was August in the Deep South and he didn't have air-conditioning, so "I left at dusk and drove all night," he says. After dropping off the fowl at dawn, he turned his car north to Washington.

Richardson proved a master in the difficult art of raising and breeding red jungle fowl. Three years later, he had expanded the flock to seventy-five. For the next three decades, he kept it healthy and isolated from other fowl to avoid diluting its genetic makeup. He gave birds to other amateur ornithologists, but most of those quickly succumbed to disease or stress. Even Beebe's New York Zoological Park in the Bronx found them too difficult to manage. Richardson had some magic touch, and his extraordinary accomplishment achieved legendary status among the small circle of people aware of the skill and devotion required to maintain these difficult birds.

Brisbin eventually returned to South Carolina and studied the rate at which chickens scratching in the radioactive Savannah River Site soil—and later, poultry exposed to Chernobyl's radioactive brew—could shed that toxicity. (He found that they do, and quickly.) He also published articles on radiocesium contamination of snakes, wood ducks, and feral hogs, and spent years researching alligators that lived in the hot-water plumes of the Savannah River's cooling plants, which earned him footage on Marlin Perkins's popular television show *Wild Kingdom*. He didn't raise red jungle fowl during those intervening decades, but Brisbin says he always remembered what Bump had told him during that 1970 phone call. "Someday," the New York ornithologist had prophetically warned, "they might be the only ones left."

A quarter century later, Brisbin noticed that a special symposium on tropical Asian birds was planned for the 1995 American Ornithologists' Union meeting in Cincinnati. "I thought, wow, here's a chance to wave the red jungle fowl flag," he says. His paper's title—"Is the Red Junglefowl One of the Most Endangered Birds in Southeast Asia?"—was designed to provoke.

The International Union for Conservation of Nature classifies three of the world's four jungle fowl as in potential danger. The exception is the red jungle fowl, which lives in much larger numbers than its sibling species across South Asia and is rare only in the crowded city-state of Singapore. Brisbin argued not that red jungle fowl were disappearing but that the wild stock was losing its genetic integrity. It was species death by introgression—the mixing of genes—rather than physical extinction.

This is not a popular issue among conservationists, since physical extinction rather than genetic introgression is the overwhelming threat to celebrity wild animals like blue whales, Siberian tigers, and polar bears as well as tens of thousands of less beloved species. The wild Muscovy duck is threatened by crossbreeding with domestic mallards, while the limited populations of wild dogs around the world increasingly mix with feral and domesticated varieties. Plants also face challenges; wild rice strains, for example, are dying out across Asia. Brisbin and several other ecologists point out that chicken, duck, and rice are critical parts of humanity's food supply, and that ensuring the genetic survival of their wild ancestors therefore is an important and prudent endeavor.

"I wanted to see if anyone would jump up and argue with me," says Brisbin. His strategy worked when the biologist Town Peterson from the University of Kansas sprang up in the conference room to insist that introgression was unlikely to have a major impact on the wild bird. The two decided to collaborate to determine the truth. Since neither is a geneticist and sequencing was in its infancy, they needed a trait that provided a single clear physical difference between the wild fowl and the barnyard variety. They settled on eclipse plumage, since ornithologists knew that a full-blooded wild male sheds its red-and-yellow neck feathers and central tail plumage in late summer for a temporary ensemble of purple plumage, while chickens do not. As Beebe remarked, the presence of these purple feathers was a reliable sign that an individual bird was free of domesticated genes.

The effort took four long years. Searching through dusty drawers and musty storage rooms in nineteen museums across the United

States, Canada, and Europe, they came up with 745 red jungle fowl specimens collected over two centuries. By comparing the dates, seasons, and locations in which the specimens were collected, the scientists uncovered a distinct and disturbing trend. Eclipse feathers start disappearing from specimens dating back to the 1860s in Southeast Asia, and this seems to spread west over time. By the 1960s, as Bump was collecting his birds, eclipse plumage began to vanish almost entirely in the last western redoubts of northern India. This change, Brisbin and Peterson believe, was not simply a natural variation within the wild population. Specimen tags suggested that many of those lacking eclipse plumage were taken from areas where the domesticated chicken was plentiful. Northern and western India, where Bump had concentrated his efforts, could be the wild bird's last stronghold.

In a joint 1999 paper, the two researchers warned that the "genetically pure wild-type populations may be severely threatened," and that red jungle fowl studies based on existing stocks in zoos or in the wild "are likely to be tainted with domestic genes." That called into question decades of studies comparing the wild fowl with chickens in order to flesh out how, when, where, and why the chicken was domesticated. Alarmingly, the red jungle fowl, "so important economically and culturally to humans, is apparently in danger of genetic extinction."

Leggette Johnson's farm on Gold Finch Road in Cobbtown sits amid the flat cotton fields of northeast Georgia, a two-hour drive south of Brisbin's South Carolina home. On one side of the modest house is a spacious fenced-in area full of long pens taller than a man. Johnson meets us at the gate under an overcast autumn sky. With the wary look and the paunch, drawl, and overalls of the quintessential Southern good old boy, he is one of the few people in the United States successfully raising jungle fowl, and his flock includes descendants of the birds collected by Bump and cared for by Richardson.

"You go in there and they'll go crazy," he says, pointing a stubby finger at a large wire pen. It is a statement, a warning, and a dare. Three birds are edging nervously to a far corner. One sports what looks like a little white turban above its small, dull brown body. Dental floss, Johnson explains. A hawk swooped down a month ago. Though the wire foiled the predator, the intended prey—a female red jungle fowl—shot against the cage in an instinctual attempt to flee. Her head split open, so Johnson retrieved some floss from the bathroom medicine cabinet, grabbed the dazed bird, and sewed up her wound while sitting on an upturned white plastic feed bucket.

Escape is always on their minds. Johnson points at the largest of the three birds huddled together behind the wire a dozen feet away. Against the monotone tan of the two hens, the rooster is a showy mass of blue and red and yellow that glows in the overcast. He bolted one day when the cage door was not shut quickly enough and was on the loose for three months, remaining near his hens but long foiling attempts at capture. "I couldn't come anywhere close," Johnson says. "If I did, he would haul ass." The fowl trusted no human, save for a neighbor's two-year-old boy in diapers, who could walk right up to the liberated animal but obviously posed no threat to the rooster's freedom.

To get to the pen with the red jungle fowl, Johnson wades through an adjacent cage with other equally wild birds. Elegantly tailed pheasants and plump quail scatter as we walk through, more confused than distressed. "These I can feed out of my hand," he says as they dart around our feet. "But not those," he adds, pointing at the three huddled red jungle fowl. "People think you're crazy when you tell them this, but if you get them really excited and catch one, it will quit fighting and limber up. Dies of heart failure, I guess."

On closer inspection, the brown of the two hens has a reddish hue with delicate black stipples on the neck. Their beaks are tiny, and they lack the spurs and combs and wattles of the foppish male. I decline Johnson's offer to let me accompany him into their cage, since I don't want to be responsible for giving such rare specimens a heart attack. There are only a hundred or so of this strain left on earth.

Johnson shrugs, adjusts his cap, undoes the latch, and steps gingerly inside. An explosion of blurred wings alters the air pressure and

I involuntarily jump. When the farmer exits the coop moments later, the birds are huddled even tighter against the far corner in a posture that seems to mix abject terror with haughty resentment. When I ask him how this bird became the domesticated chicken, he doesn't answer but leads me to the other side of the yard.

The Georgia farmer has one of the few collections of all four jungle fowl in the United States. The females all tend to be plain and brown and lack combs, the better to avoid detection when incubating their eggs on the forest floor. The roosters, splashed with vibrant colors, are more dazzling to bird eyes, which have four color cones to our three. Darwin explained such extravagance as an arms race among males in order to appear more appealing to their potential mates. Scientists now say that they are also trying to impress their competitors. Like the plumed helmets of ancient Greek warriors or the bright pantaloons and turbans of the nineteenth-century Zouave soldiers, livery can dazzle and psych out the enemy.

Johnson leads us first to a cage containing the Sri Lankan jungle fowl native to that teardrop-shaped island off the southeast coast of India. The rooster and hen move cautiously to the back of the pen, but they don't panic. The male is similar in size and shape to the red, but with a yellow-orange palette with a splash of yellow in the comb. The next is the gray jungle fowl of southern India, and the rooster races back and forth in its cage on black legs, rustling its black-and-ocher feathers set on a grayish background with a bit of yellow on the neck. This hen, like the others, is plain, but has yellowish legs.

In the next pen is a green jungle fowl. Its natural home is on Java and Bali, islands of today's Indonesia, more than two thousand miles east of Sri Lanka. This rooster stands strangely immobile, and stares at us with an unnerving intensity. He seems confident in his magnificent plumage, which is the most dramatic of all jungle fowl. His body shades from the color of long-exposed bronze to an emerald green. Feathers at his throat are a sky blue and bright violet, with splashes of ocher and electric yellow, and his comb shades from light blue to deep red.

What's odd, Johnson says as we stand in front of the motionless green jungle fowl, is that these three sister species are skittish, but not suicidally so. He's never had to use dental floss on the birds on this side of his barnyard. Nor do the grouse, quail, partridge, and golden pheasant in the other pens approach the wild and untamed spirit of the red jungle fowl. He tries to avoid going into the pens with the reds more than once in three days to limit their trauma.

The unusual nature of these red jungle fowl recently drew the attention of Leif Andersson, a biologist at Sweden's Uppsala University who has pioneered DNA sequencing of domesticated animals as a way to track their genetic past. He was part of the team that published the chicken genome in 2004. As Brisbin had decades earlier, Andersson realized that he needed a reliably wild chicken to compare with the domesticated variety in order to map the differences more accurately. In 2011, he visited Johnson's farm to take a look at these unusual animals and sample their blood. The DNA of the Richardson birds, now being sequenced in Andersson's Uppsala lab, could help unlock important clues to the chickens' murky history, particularly if they prove to be among the last with an undiluted genome.

On our way back to South Carolina, Brisbin muses on the mystery of chicken domestication. Biologists are still arguing over when, where, and why the bird left the jungle for the backyard. Thousands of years ago, somewhere in South Asia, it merged with human society. Our farming ancestors may have welcomed an animal that feasted on weeds and pests, hunters may have captured it in the forest and brought home live birds that eventually were tamed, and foragers could have found unhatched eggs and incubated them artificially. Brisbin believes, however, that only a genetic mutation that turned off the fowl's natural skittishness could have paved the way for the more placid modern bird. "There's a five percent chance that when you hold one it will die," he says of Johnson's jungle fowl. This shift from wild creature to barnyard chicken—possibly just a random turn of the genetic dials—was a dramatic transformation of the animal, with big implications for our own species.

Suddenly, a squirrel darts in front of the car. I swerve, but hit the unfortunate animal. Brisbin orders me to turn around. "Do you have a bag?" he asks with an almost childlike eagerness. "Let's not waste it." We return to find its head flattened, but it is otherwise unmarked. "It's perfect," he says, placing the bag in my backseat. Then he chuckles and confides that his collecting unnerves the sentries at the gate of the Savannah River nuclear site. "The guards don't want to check my car because there may be a snake or an alligator in it." He glances into the backseat. "Remind me to retrieve it when we get back to my place."

A couple of months later, I call the retired Richardson, curious as to how this self-taught slaughterhouse owner maintained Bump's difficult birds so successfully for so many decades, despite the ever-present threat of heart failure or disease. If anyone would have a gut feeling as to how this bird became the docile creature of today's factory farms, it would be him.

A woman answers the phone in his home in Tuscaloosa. "I buried him six weeks ago," Richardson's wife says. "He was eighty-three years old and had never been in the hospital." I give her my condolences, and then her daughter takes the phone and introduces herself. "He was very particular about their care," she explains when I ask about the red jungle fowl. "He kept them separated and wouldn't let them breed with other birds. And if anyone else came around—even me— the birds would be all in a tear." She adds that just a couple of weeks before he died, she asked him why he prized these fussy and difficult fowl so highly. "He said, 'I like them because there is no way to tame them.' He said, 'I like them because of what they are—wild.'"

2.

The Carnelian Beard

And lo, the Rooster King, how he slums like the Lord!
And lo, the Rooster King, how he chases from these vacant
   lots the lesser, more domestic, cocks!
And lo, the Rooster King, how he spreads, as gasoline,
His wings, O, stained-glass butterfly!

     —Jay Hopler, "The Rooster King"

The chicken's grandest entrance in history took place on a bright autumn day in 1474 BC, when four fowl were carried triumphantly into Thebes, the world's largest and wealthiest city, under the eyes of the powerful Egyptian pharaoh Thutmose III. The whole city turned out to see the lavish procession. Magnificent horses pulled chariots plated in gold and electrum that reflected the sun, blinding the crowd as they rolled down the city's triumphal avenue. Nubian retainers carried huge ivory tusks from Syrian elephants shot by the pharaoh himself. Strangely clothed captives singing in guttural tongues passed by, as did prancing bears and a live elephant—all loot from a six-month-long military campaign in the Middle East. The sons of foreign princes who were kept as hostages trundled by in gilded cages similar to those that held the colorful birds.

The four exotics were tribute paid by Babylonian princes to the king, who is still remembered as one of Egypt's most successful military and political leaders. Only five foot three inches tall, Thutmose III's conquests took him across the Euphrates River and into Mesopotamia—today's Iraq—where he had thrashed his opponents and returned to his capital city that Egyptians called Waset and Homer praised as "the world's great treasure house . . . with its one-hundred gates."

Thutmose had the annals of the successful invasion carved into the walls of Thebes's great temple of Karnak, which lay at one end of the wide avenue that led to the capital's Luxor sanctuary. Since there was no hieroglyph to describe this previously unknown creature and the stone is damaged at key spots, there is no absolute certainty that the gifts were chickens. But the inscription identifies a Mesopotamian bird that does something every day, which Egyptologists think is laying an egg. Not until the twentieth century were most hens capable of laying an egg a day, so that claim may have been among the many boastful exaggerations carved into the blocks. But no other bird fits the description as well as the chicken.

Ancient Egypt was for much of its history an insular land defined by its life-giving Nile River and its fertile delta that pushes into the Mediterranean Sea. With rich harvests, deserts to the east and west, the sea to the north, and the African savannas to the south, Egyptians for much of their early history did not stray far from their valley. That changed during the early New Kingdom, when Thutmose III's grandfather Thutmose I moved troops into the Levant. They were mesmerized when they experienced rainfall for the first time, calling it "the Nile falling from the sky." The heyday of the New Kingdom brought the world to Egypt's rulers. Thutmose III's stepmother, the pharaoh Hatshepsut, sent ships down the Red Sea to the land of Punt, which may have been today's Somalia or Ethiopia, to obtain frankincense, ivory, ebony, and myrrh trees to transplant in Egyptian soil. She also encouraged international trade in the Mediterranean. The chicken's arrival marked more than the spread of one species. It underscored Egypt's newfound interest in exploring and conquering the outside world and collecting foreign goods, plants, and animals.

Thutmose III was personally entranced by exotics. Near the Karnak walls inscribed with the mention of the bird that lays every day is a room with carved representations of pomegranates, iris, gazelles, and other species not native to Egypt. Across the Nile, on the barren west bank set aside for tombs and funerary temples, his vizier Rekhmire was buried with metal vases thought to be a tribute brought by Minoans from the island of Crete. One is incised with images of lion, bull, and antelope heads. There is also a crudely drawn bird with two wattles, a comb, and a pointed beak that may be one of the oldest known images of a rooster.

The chicken was a rare and royal bird in ancient Egypt, a fact that only came to light in 1923. Working in the Valley of the Kings, west of Thebes, Howard Carter unearthed the tomb of King Tutankhamen in late 1922. Four months later he reported finding a broken piece of pottery between the nearby tombs of Ramses IX and Akhenaten, Tutankhamen's predecessor and a religious renegade who ruled a century and a half after Thutmose III and focused Egyptian worship on the sun god. Carter was overwhelmed with the thousands of objects stuffed in King Tut's burial chambers that would consume the next decade of his life, as well as the immense global publicity surrounding the find. His sponsor and friend Lord Carnarvon had died just two weeks before in Cairo—either from an infected mosquito bite or, more popularly, a curse placed on the tomb. Yet the sought-after Egyptologist was excited enough by the little pottery piece to write an extended paper of what he called not only "the earliest known drawing of the domestic cock in the form of the Red Jungle-fowl" but "absolute authentic evidence of the domestic fowl . . . being known to ancient Thebans."

The drawing on the little triangular-shaped bit of limestone, now kept in the British Museum, is utterly charming. It lacks the formality and stilted quality of so many ancient Egyptian statues and friezes. Instead, it is a bold ink sketch with a large serrated comb, prominent wattles, closed wings, and a wide flared tail. This rooster is clearly drawn by someone who has seen a live one. It struts its stuff like any barnyard fowl, "which suggests," adds Carter, "that in that early period its domestication was already accomplished."

Based on its location, he dated the small fragment to between the time of Akhenaten, around 1300 BC, and the end of the New Kingdom in 1100 BC after a series of devastating droughts and wars left it at the mercy of foreign rulers. This is only a rough guess, but Carter's find was long the oldest indisputable image of a chicken in Egypt. The other candidate is a silver bowl that shows a rooster amid scenes of farming, an elegant vessel found in the Nile delta with intriguing elements of Egyptian and West Asian styles. Thought to date to about 1000 BC, the vessel in fact may have been made in the glory days of Ramses II in the thirteenth century BC, according to the art historian Christine Lilyquist. The rooster's presence on the artifact may hint at a religious rather than an agricultural role for the bird, since it was found in a temple to the cat goddess Bastet, who protected Egyptians from contagious diseases and evil spirits.

But if the chicken was special, it remained alien. In the unsettled centuries following the demise of the New Kingdom, Egyptians buried more than rulers and viziers. They also mummified hundreds of thousands of animals. From cats to crocodiles, the dead creatures were preserved for the afterlife because they were dearly beloved pets, to feed the human dead, or in honor of a specific god in the complex Egyptian pantheon. Amid more than thirty cemeteries that hold 20 million animal remains, including 4 million ibis, archaeologists have yet to find a single mummified chicken. The exotic in ancient Egypt was fascinating, noteworthy, and a sign of status. But it was not sacred.

By contrast, there are lots of goose mummies. When the chicken arrived in Thebes amid much pageantry, geese were already domesticated and living in poultry yards. A migratory bird, they were at first seasonal visitors to the Nile valley. Drawn to the rich grain fields of ancient Egypt, they learned to live with humans, providing a ready supply of meat, a couple dozen large eggs a year, and an effective warning system against intruders.

In the long run, geese proved no match for a bird that could be bred to lay more eggs, grow much faster, and eat a wider variety of food, including the ticks and mosquitoes that flourish in the humid Nile environment. Roosters also were reliable heralds of dawn in the era before alarm clocks. In a land of farmers, this was a welcome trait. The new arrival was poised to overtake its competitors as the most useful bird in the West.

Twenty-five hundred miles separate the Nile valley from the western-most edge of the red jungle fowl's habitat in Pakistan. The chicken's journey from east to west coincided with the rise of the world's first three urban civilizations a thousand years before Thutmose III's reign. Thanks to bones, clay tablets, and a scattering of artifacts, ar-chaeologists have begun to trace how the bird leapfrogged from one culture to the other in the first stirrings of a globalized world.

That journey began along the two-thousand-mile-long Indus River, which flows out of the Himalayas and empties into the Ara-bian Sea. While the Egyptians of the Old Kingdom labored on their pyramids, the people of the Indus civilization built a society larger and more populous than either Egypt or Mesopotamia. Its half-dozen major cities boasted broad streets, innovative water systems, and sewers unmatched in the ancient world until the rise of Rome. A system of symbols—as yet undeciphered—was widely used. Wheeled carts, riverboats, and seagoing ships connected an area exceeding a million square miles with a population of several million people, who grew barley, millet, wheat, and rice in fields plowed by water buffalo; herded goats and sheep; and hunted boar and wild birds.

Indus hunters would have been familiar with the red jungle fowl on the northern end of their civilization in the Himalayan foothills, as well as the gray that lived on the far southeast fringe, around the Indian province of Gujarat, which touches the Arabian Sea. Just a few days' walk from the red jungle fowl's present home in the Himalayan foothills is the northern metropolis of the Indus, the city of Harappa. Harappa once sprawled over two hundred fifty acres along the Ravi

River, its walled neighborhoods home to twenty-five thousand people or more during the Indus heyday between 2600 BC and 1900 BC—roughly contemporary with the second half of Egypt's Old Kingdom.

Unfortunately, Harappa's first excavators were British railroad engineers in the middle of the nineteenth century. Surprised and delighted to find thousands of buried baked bricks in a rural backwater, the engineers directed their local workers to use this material as the bed of the Lahore railway. At the time, no one suspected that an Indus civilization ever existed. Only later, as Carter was examining the piece of Egyptian pottery depicting a rooster, did archaeologists realize that Indian passengers and freight were clicking along on tracks that sat on the reused scraps of one of the world's first great cities.

Richard Meadow and his colleague Ajita Patel worked for years at Harappa until Pakistan's uncertain security situation halted excavations, and they have the best collection of bird bones from an Indus site. On a cold spring day I visit them at their zooarchaeology laboratory at Harvard University's Peabody Museum in Cambridge. The lab's entrance is just to the right of a Mesoamerican mural of a warrior holding his rival's severed head. Meadow, a tall and taciturn New Englander, and Patel, a diminutive and talkative Indian woman, share a crowded office piled high with books and papers and decorated with eerie wooden masks made by a Creek Native American.

They take turns explaining that identifying chicken bones is no simple task. Many archaeologists fudge by giving them a generic label such as chickenlike or chicken-sized. The region is rife with cousins to the domesticated bird, such as the chukar, francolin, and quail, which leave behind similar bones. Thousands of years of wear and tear make identification even more difficult. Differentiating a red jungle fowl bone from that of a chicken is even more problematic. Though today's domesticated bird is generally much larger than the wild ancestor, this may not have applied four millennia ago.

And if chickens were an important source of food, Meadow adds that the Indus people may have eaten much of the evidence by crunching on the ends of bones, a practice he says is still common in that part of the world. These cartilage structures provide the best clues to a particular species. Extracting DNA from Indus bird bones holds promise, but many of the bird remains were excavated years ago and have been subject to decades of possible contamination.

Harappa is close to the red jungle fowl's current range, but other Indus sites are not. At Mohenjo Daro, located halfway between the mountains and the sea in Pakistan, scientists found a four-inch-long femur bone that excavators describe as very "chickenlike." Today's modern factory-farmed chicken has a femur measuring just shy of five inches, while the red jungle fowl's averages less than three inches. West of Delhi, the Indian archaeologist Vasant Shinde has discovered similar bones in what was a modest Indus town that he is convinced come from chickens.

Along with a dearth of chicken bones, the Indus people left behind frustratingly few traces of their day-to-day life in this enigmatic civilization. Scholars have yet to decipher their symbols, and unlike ancient Egyptians and Mesopotamians, they produced no known life-sized statues, only tiny figurines. One enterprising archaeologist recently sorted through thousands of small hand-molded clay figurines found at dozens of Indus excavations and largely neglected by previous generations of researchers. Among the figures in various collections in Harappa, the Lahore Museum, and the National Museum, New Delhi, were several that are decidedly chickenlike. One seems to sport a comb and a curved tail similar to that of either a chicken or jungle fowl. Another looks like a rooster wearing a collar. That doesn't necessarily mean it was domesticated, since wild birds might be chained, and other figurines of rhinos and tigers occasionally have collars as well. Two birds feeding in a dish may or may not be chickens, but the scene is a compelling image of domesticated birds. The Indus people did keep birds in cages. A small terra-cotta cage was recently unearthed that is similar in size to those still used in modern Pakistan for housing live partridge and quail.

One of the most striking figures is a clay man holding a fowl-like bird calmly in his arms, up against his chest, as men on the subcontinent still do before a cockfight. A rooster spur found at Harappa and a clay seal with what may be two roosters facing each other are circumstantial evidence that cockfighting, still popular in India and Pakistan today, was practiced four millennia ago. Some southern Indian traditions combine cockfighting with religious rituals associated with a mother goddess that may be of ancient origin.

Strict Hindus forbid meat eating, and the specific taboo on the bird may stretch back to a time when the animal, like the cow, was considered especially sacred. But recent analysis of Indus cookware shows that they had most of the ingredients they needed for a good chicken curry, a term that likely derives from *kari*, the word for "sauce" in the South Indian language of Tamil. Baffled by that region's wide variety of savory dishes, seventeenth-century British traders lumped them all under the term *curry*. A curry, as the Brits defined it, was a mélange of onion, ginger, turmeric, garlic, pepper, chilies, coriander, cumin, and other spices cooked with shellfish, meat, or vegetables. But no one knew how old curry might be.

Working with other Indian and American archaeologists, the archaeologist Arunima Kashyap, then at Washington State University in Vancouver, applied novel methods for pinpointing the elusive remains from old cooking pots found at Shinde's site. They also obtained human teeth from a nearby cemetery dating back to the same era. Back in her lab Kashyap examined the samples using a technique called starch grain analysis. Starch is the main way that plants store energy, and tiny amounts of it can remain long after the plant itself has deteriorated. If a plant was heated—cooked in one of the tandoori-style ovens often found at Indus sites, for example—then its tiny microscopic remains can be identified, since each plant species leaves its own specific molecular signature. To a layperson peering through a microscope, those remains look like random blobs. But to a careful researcher, they tell the story of what a cook dropped into the dinner pot forty-five hundred years ago.

Examining the human teeth and the pot residue, Kashyap spotted the telltale signs of turmeric and ginger, two key ingredients of a typical curry. Wanting to be sure, she and a colleague abandoned the lab for her kitchen. Using traditional recipes, they cooked the dishes and then examined the residues to see how they broke down. The results matched what they had unearthed in the field, confirming that they had found the oldest examples of ginger and turmeric and the first spices from the Indus era. Ancient cow teeth from Harappa also showed signs of ginger and turmeric. The Indus people may have done what villagers still do and placed leftovers outside their homes for wandering cows to munch on.

And what would curry be without a side of rice? Archaeologists once thought that Indus farmers were restricted to a few grains like wheat and barley. Working with colleagues at two ancient sites near Delhi, the Cambridge University archaeologist Jennifer Bates found remains of rice, lentils, and mung beans. The rice discovery was a particular surprise, since the grain was long thought to have arrived only at the end of the Indus civilization. In fact, inhabitants of one village appear to have preferred rice to wheat and barley, though millet was their favorite grain. Shinde thinks that all the important ingredients were there for one of the most common dishes today in any Indian restaurant, and other archaeologists suspect that many of the traditions developed in the Indus era—religious, social, and technological—continued in later Indian civilizations, including tandoori chicken.

With all its exotic ingredients, curry took thousands of years to catch on in the Middle East and Europe, but the adaptable chicken was poised during this age of the first great civilizations to make its first leap to the West. At an Indus site called Lothal on the western edge of India, archaeologists uncovered chickenlike bones and personal seals owned by merchants who lived along the far-off Persian Gulf. In the middle of town, now excavated, is an enormous brick-lined reservoir, which many researchers believe was an artificial harbor. Nearby were warehouses, a bead factory, and a metalworking area.

Lothal was once a thriving center of the earliest oceangoing trade. From here, sailors could strike out across the Arabian Sea and cover a thousand miles of open ocean using the monsoon winds off the coast of Arabia, and then work their way up the Persian Gulf to the busy wharves of the Mesopotamian metropolis of Ur, then the largest, wealthiest, and most cosmopolitan city on earth.

↞ ↞

The merchant paces back and forth impatiently on the busy quay. His large wooden ship, sails furled, is tied to the dock as a government bureaucrat methodically records every item that the sweating stevedores bring up from the packed hold. The smell of local mutton from the nearby food stalls mixes with the aroma of exotic spices. Between loads, the clerk glances at a bird in the wicker cage in the warehouse shade. "What do we call that?" he asks. The merchant shrugs—he knows nothing about birds—and the bureaucrat uses his pointed reed to impress symbols on a wet clay tablet as the next case of goods is set down. Once the documentation is complete, the merchant's first stop is the royal palace on a high mound above the harbor of Mesopotamia's great city of Ur. King Ibbi-Sin has a pleasure garden full of exotic animals, and he is sure to be pleased by the richly colored bird.

According to Genesis, Abraham, the legendary father of the Israelites, left his home and family in Ur for the greener pastures of Canaan at about this time. He was an exception. The city in 2000 BC drew traders from distant lands and women from local villages looking for work in its busy textile mills. With its huge temples and palaces and its busy wharves on the Euphrates River, which leads to the Persian Gulf, Ur lay at the center of a prosperous kingdom controlling a large part of what is now southern and central Iraq. The merchants here were the first to use money in the novel form of silver shekels rather than traditional but cumbersome units of grain. Scribes recorded even the smallest of transactions by etching cuneiform signs on damp clay tablets. The dynasty's founder, Ur-Nammu, created the world's first formal legal system, and his son Shulgi, who succeeded him, was not only literate—a rarity for any ruler at the time—but revised the curriculum of the scribal schools, built roads, and provided the first inns for travelers. Shulgi also is credited with creating the world's first zoo by collecting exotic animals from far-flung lands.

The royal pastime of collecting animals like camels and oryx that were unknown to Mesopotamia continued for several decades until the reign of the dynasty's last king, Ibbi-Sin. The king of Marhashi—likely part of today's Iran—sent him what was reported by puzzled scribes as an extraordinary speckled dog that might have been a leopard or hyena. We know this thanks to the thorough bureaucrats of Ur, who in the space of little more than a century left behind more than a hundred thousand clay tablets. One tablet, dated to the thirteenth year of Ibbi-Sin's reign, mentions a bird of Meluhha among a list of other items at the palace at Ur. It is one of five known references in the archive to this creature. Some may have referred to live birds, and others to statues or curios made of wood or ivory.

The word used for the bird is *dar* in Akkadian, the intricate Semitic language predating Hebrew and Arabic that was used in Mesopotamia for more than twenty-five hundred years. Translating from a long-dead language to modern English is always treacherous, but scholars largely agree on the names for wrens, ducks, crows, sparrows, pigeons, and other native birds. Exotic animals, however, are much harder to identify in the historical record. Names for exotic animals can be confusing even today. A Turk invited to an American Thanksgiving might wonder why the New World main course is named for their Anatolian Old World country. In 1533, an Italian naturalist called the turkey "the wandering chicken," and a French scientist later added in the Greek name for the guinea fowl, giving us today's scientific name of *Meleagris gallopavo*. Our common name comes from European confusion over the guinea fowl's native land, which is Africa rather than Turkey.

Names are often tied to specific local varieties. A modern rancher chatting about Texas shorthorns and Aberdeen Angus knows that he means different kinds of cattle, but a Seattle vegan hipster might have no clue that the guy in a cowboy hat is referring to cows. The difficulty extends beyond types of animals. Ancient Mesopotamians perceived color differently than we do. What is called "spotted" may be "speckled" or even "red" depending on which Akkadian specialist you ask.

Some specialists think that, based on a variety of clues, a *dar* is a predominantly dark-colored bird, possibly a black francolin, a wild pheasant native to the area. It would be an obvious bird to compare the chicken with, since both are in the pheasant family and bear a strong resemblance to each other. At a time when transporting animals was expensive, difficult, and hazardous, dubbing the strange bird a "black francolin from India" made sense.

Other clues point to the chicken's arrival in Mesopotamia from the Indus. One is lodged among one of the oldest of recorded stories. Called "Enki and the World Order," this legend tells of the god of water surveying the order he has brought to creation. In the land of Meluhha, Enki praises the forests and the bulls. "And may the *dar* of the mountains wear carnelian beards!" he thunders. The Indus people used the deep-red stone called carnelian to make beads, many of which were exported to Mesopotamia through ports like Lothal, and a red-bearded bird certainly is a good description of a rooster's wattles.

Archaeologists are not likely to find any chicken bones to confirm that the bird was brought from the Indus to Mesopotamia in the first heyday of international trade, because not long after the scribes noted the arrival of the black francolin of India, tribes from the north and east swept down on Ur, sacked the metropolis, and carried off the king to captivity and then death in Iran. The catastrophe marked the end of southern Mesopotamia's control over the region. If the chickens imported from Meluhha survived, they did so in numbers far too small to show up in a dusty trench.

The skeleton mounted on a wooden stand high on a bookshelf in Joris Peters's office at a Munich university looks like a baby ostrich. Peters, a zooarchaeologist, laughs at my mistake. "No, no, the sternum in an ostrich would be flat," he explains, grabbing the model and placing it on his cluttered desk. He points at the large curved bone shaped like a boat's keel. "The chicken has a sternum with a furcula—a wishbone— that helps it fly."

Peters is a trim and clean-shaven Belgian scientist who spends much of his time at excavations in the Near East. He also oversees one of the world's largest collections of ancient chicken bones in a storage facility one floor below his office, although it takes up only a few shelves. Unlike cow, sheep, and goat bones, chicken remains usually vanish in their entirety, since humans, dogs, or other scavengers typically make short work of a carcass.

While Peters worked on a dig in Jordan in the 1980s, his team ate a chicken a day and dumped the scraps just beyond the camp. He noticed that each night, predators made off with nearly every bit of the carcass. Curious, he checked each morning to see what was left behind. When the season was over, he calculated that only a single bone a year would survive in that desert environment. In wetter environments, where bone can degrade more quickly, the odds are even lower.

Until about twenty years ago, most archaeologists did not bother to save bird bones, which were not considered very interesting or significant at digs. Researchers now realize that these remains provide important clues to diet, social organization, trading patterns, and the state of the environment thousands of years ago. With a cheap fine-meshed screen, archaeologists now are finding hordes of tiny bird-bone fragments. Yet even when they survive the march of time, chicken bones can be as hard to interpret as Akkadian nomenclature. A francolin and a red jungle fowl femur look similar. And buried chicken bones behave differently than do the heavy ones of, say, a slaughtered sheep. They can slip lower in the soil and into older archaeological layers, and rodents digging underground can easily shift them about.

Used chicken bones also are treated differently from those of cattle or sheep. Large bones might be buried on a settlement's outskirts, while bits of chicken could simply be tossed away near dwellings. Over time, as buildings were torn down, repaired, and constructed, these remains could find their ways into later structures. A bone found next to a pot made in the time of Caesar might actually have come from a dinner served before Rome was founded.

Peters rifles through the large, plump freezer-sized plastic bags that cover half the surface area of his desk, opens one, and pulls out a smaller bag, then lays a half-dozen pale chicken bones next to a plate of chocolate eggs left over from Easter. These little bits of cartilage are, at the moment, the oldest known physical evidence of chickens in the Near East or Europe. In Peters's collection since the 1960s, they were excavated from an ancient settlement in Turkey. Based on the layers in which they were found, they date to between 1400 BC and 1200 BC, about the time the chicken made its debut in Egypt.

The day before, Peters mailed four or five of the thirty-odd bones found at this Turkish site to a British colleague, who will individually date each bone using radiocarbon techniques and try to extract enough of the animal protein collagen from inside the bone to capture the ancient bird's genetic sequence. The collaboration is part of Peters's ambitious effort to follow the chicken's spread across Asia and into Europe using more precise dates and ancient DNA. That means laboriously examining hundreds and even thousands of bones sealed up in little plastic bags that may or may not still contain DNA.

Scientists like Meadow, Patel, and Peters have reason to be extra cautious in interpreting old chicken bones. In 1988, for example, a Chinese and a British archaeologist reported finding an eight-thousand-year-old chicken bone in central China, more than a thousand miles to the north of the red jungle fowl's habitat. The news made global headlines, since the bone was twice as old as the chickenlike ones gathered in the Indus. The oldest textual evidence for Chinese chickens dates to about 1400 BC, the same era as its arrival in Egypt and Peters's Turkish sample.

If true, the discovery meant that chicken domestication likely took place long before agriculture emerged in the region and that it spread quickly out of the wild bird's habitat and into chilly northern latitudes. In this scenario, the early chicken may have worked its way east from northern China across Russia to Europe, bypassing India and the Middle East altogether. The specimen appeared to revolutionize our understanding of how humans and animals moved around in prehistoric times.

Peters recently went to China and examined the bone, however, and determined that it likely is no more than two thousand years old and worked its way from more recent to older layers. The archaeologists dated the layer, but not the bone itself. Other Chinese finds labeled as ancient chicken have turned out to be partridge. Until there are better data from old bones, the chicken's path north from South Asia into China and then on to Korea and Japan remains theoretical.

Meadow and Peters themselves have nearly been fooled. When Meadow worked as a young student at a site in southeastern Iran, the team found a chicken-leg bone complete with spur from what seemed to be a large domestic rooster. The layer of the ancient settlement dated to 5500 BC, a full millennium and a half before the Indus civilization flourished a few hundred miles to the east. Were it that old, the bird's presence suggested that the chicken already was outside the red jungle fowl's range long before the emergence of the Indus and Mesopotamian civilizations. Meadow noted, however, that the bone was bleached whiter than other animal remains in that layer, and prudently concluded that it probably came from the first millennium BC and had worked its way to the older level. In 1984, Peters found a chicken bone in Jordan that appeared to date from the time of ancient Ur, around 2000 BC. In the early twenty-first century, an American archaeologist working at the same site found a similar bone in material dating to the same era, but radiocarbon dating of both samples showed the remnants came from a medieval chicken dinner.

Just beyond a military checkpoint, backed up against the high and rocky hills of northern Iraq, is the village of Lalish. The steep mountain road that leads to the small town eventually narrows to the width of a narrow street, and visitors remove their shoes before proceeding further uphill between tall stone buildings. Lalish is the sacred center of the Yezidi, an embattled religious minority in a country riven with sectarian strife. "We are the oldest religion on earth," explains Baba Chawish as he offers me a seat on a velvet sofa while he folds his long limbs gracefully onto a cushion on the floor of the small reception room.

The Yezidi priest is a striking man, tall, with a long nut-brown face and a thick, dark beard below a flattened pale turban around a black skullcap. A black sash sets off his white robes and cream vest; even his cell phone is an elegant white. Long persecuted by Christians and

Muslims, the Yezidi revere an archangel named Tawûsê Melek who refused to bow down to any being but God. Their antagonists identify him as Satan, and accuse the Yezidi of devil worship. Scholars say that the religion has ancient roots that predate the Abrahamic faiths and has since absorbed a host of later traditions.

Tawûsê Melek takes the form of a peacock, another exotic bird from the East that was brought as early as 2000 BC to Ur, which lies five hundred miles to the south. According to Yezidi beliefs, the sacred peacock landed in Lalish and then met Adam in the Garden of Eden to instruct him on solar worship. The rooster is also held in high esteem. "He tells us when to pray," the baba explains, and I notice a small stuffed cock standing on top of a clock in one corner of the room. Pious Yezidis face the sun five times a day to recite their prayers, and the cock's crow before dawn signals the start of the daily rite.

The earliest evidence for the chicken's role in religion was found less than one hundred miles to the south, along the Tigris River amid the ruins of the ancient Assyrian capital of Assur, just upstream from Saddam Hussein's hometown of Tikrit. A massive ziggurat, the stepped pyramid favored by Mesopotamians, towers above the valley, though its straight edges have melted over the millennia, forming a conical hill. Mounds of toppled temples and palaces rise in shorter clumps over vaulted underground tombs that have long held generations of royalty. Within one of those graves, beside the skull of a woman, German archaeologists found a delicate ivory box along with a matching ivory comb, gold beads, earrings, and a seal made from lapis lazuli mined in Afghanistan. Incised on the box is the oldest known image of the chicken on its home continent of Asia.

In the Old Testament, Assyrians are likened to wolves, while historians today disparage them as heartless conquerors. We know that at the height of their empire in the eighth century BC they forced the migration of entire populations and ruthlessly suppressed their enemies, common practices at the time. Much of the exceptionally bad press comes from their effective propaganda carved in fearsome stone reliefs while they dominated the Near East for two centuries, until their total defeat in the seventh century BC. But for most of its long history, Assyria was a small and tight-knit merchant kingdom that used its central position between southern Mesopotamia and Turkey to the north, the Levant to the west, and Persia to the east for economic advantage. Assur was the spiritual heart of Assyria, as Lalish is for Yezidis today.

The little cylindrical box, a mere three inches high, dates to the late fourteenth century BC, a century or so after Thutmose III's invasion. The birds mentioned in the pharaoh's annals may have come from this area, and the box may be slightly older than Carter's potsherd and the silver bowl of the delta. Unlike the bloody scenes immortalized later in Assyria, the box pictures a stylized Garden of Eden, a peaceful setting of gazelles grazing under palm and conifer trees with cocks and hens on the branches. Between each set of trees, a sun blazes. "In addition to being exotic creatures, they may have had some magical or ritual significance related to the new day or their fertility," says art historian Joan Aruz.

The bird, once a novel gift in Western courts, had taken on divine attributes. Assyrians worshipped Shamash, a sun god portrayed as a flaming disk, among their deities. He was the son of Sin, the moon god, and the power of light over the evil of darkness. A temple dedicated to both gods was built around 1500 BC in Assur, and one had long stood in Ur. After this brief glimpse of the bird, several centuries pass before it reappears in Babylon, which lay on the Mesopotamian plain between Ur and Assur.

Babylon in the sixth century BC was at its zenith. With help from the Medes and Persians to the east, Babylonians had destroyed Assur, vanquished the Assyrian Empire, and reasserted themselves as the center of power on the vast Mesopotamian plain. The multicolored Etemenanki—the seven-stepped ziggurat immortalized in the Bible as the Tower of Babel—stood as high as the Statue of Liberty in the center of a vast metropolis housing more than two hundred thousand people from all over the Middle East, at the time the largest urban center ever constructed. Marduk had been the patron deity and preeminent god of Babylon for more than a thousand years, but his popularity began to slip during and after Assyria's collapse as the sun and moon gods gained status.

The last king of Babylon, Nabonidus, who came to power in 556 BC and who may have been of Assyrian origin, accelerated this trend. While Egypt's Akhenaten focused on one solar deity, Nabonidus paid particular homage to Sin, the same primary deity of ancient Ur. He also worshipped the moon god's child Nusku, who symbolized light and fire and was associated with the rooster. Babylonian inscriptions of the era, written in the same cuneiform used by the Ur scribes fifteen hundred years earlier, mention the *tarlugallu*, translated as the "royal bird," which some scholars suspect was the chicken. The bird also shows up suddenly and repeatedly in practical and common objects. Ancient Mesopotamians often carried a little stone cylinder on a cord around their necks. Each was engraved with scenes of gods, heroes, and animals that became visible when rolled across a bit of clay, and served as a personal signature or mark of an institution. Several cylinder seals from this era show roosters, perched on elaborate columns reserved for sacred symbols or servants of deities, accepting the offerings of adoring male priests. Often the crescent moon hovers nearby.

Nabonidus spent fifteen years living at an Arabian oasis, far from the bustling capital. Historians still debate his motives for decamping to the desert, but his absence and his religious ideas likely angered the priests of the traditional cults, unsettled the aristocracy, and disturbed the army. When he returned, Persians and Medes—former allies in the destruction of Assur—crossed the Tigris River, won over some disgruntled Babylonian generals, and defeated Nabonidus's forces just north of today's Baghdad. On October 29, 539 BC, the famous gates of Babylon, decorated with their blue-and-gold glazed lions and bulls, were thrown open and the wide streets strewn with green reeds and palms. Nabonidus was captured by the invaders.

The entry of the Persian conqueror Cyrus the Great into the world's greatest city marked the beginning of the chicken's sudden spread throughout West Asia and into Europe. His successors eventually controlled all the lands between the Indus River and the Nile, right up the Bosporus, which separates Europe from Asia. They granted a measure of self-government to this multiethnic society, modernized the creaky old administration of Babylonia, and were careful not to interfere with religious freedom throughout the sprawling realm.

No ancient people, except possibly the Romans, would grant the chicken a greater role and higher status than the Persians and their Zoroastrian religion. "The cock is created to oppose the demons and sorcerers," states a Zoroastrian tradition. "And when it crows, it keeps misfortune away from . . . creation." The Persians held the rooster in such esteem that it was forbidden, as it was among Hindus, to eat the bird. It banished the sloth-demon Bushyasta, "who desires to keep people wrapped in slumber, even after the morning has dawned upon the earth," as one commentator puts it. The bird landed "the death blow to the world of idleness," as anyone who has attempted to sleep late in any rural area in South Asia quickly learns.

The sacred and royal nature of the cock may even have inspired one of the oldest symbols of kingship, the crenellated crown. Persian kings were the first to introduce that peculiar headgear, which remains in fashion among royalty. There are no contemporary explanations for the pointy bits on a circular diadem, and they may symbolize a castle wall, high mountains, or the rays of the sun. But the triangular shapes on the classic royal crown also resemble a cock's comb. Intriguingly, stone reliefs at the Persian capital of Persepolis include images of a crowned man with wings under a crescent moon. Another Persian sacred or royal hat, the *kurbasia*, was explicitly designed to resemble a cock's comb.

The chicken arrived in Persia, today called Iran, between 1200 BC and 600 BC, also the range of dates given for the birth of Zoroaster. According to some traditions, he was born in Afghanistan, between Iran and Pakistan. Like Jesus and Muhammad, he was called, as a middle-aged man, to reveal a new truth, overturn old traditions, and endure criticism from the clerical establishment. Zoroaster, some scholars say, sought to reform the old Iranian pantheon and elevated Ahura Mazda—a Persian deity with a name translated as light-wisdom—to the role of omniscient, omnipotent, and uncreated god.

Ahura Mazda created Angra Mainyu, a Satan-like figure who was the root of all sin and suffering and would at the end of time be destroyed. Like the Jewish Yahweh, Ahura Mazda was typically not represented in any statue or carving. And among Ahura Mazda's assistants—immortals similar to the Yezidi, Jewish, Christian, and Islamic archangels—Sraosha opposes all evil while spreading the Zoroastrian gospel of good thoughts, good words, and good deeds. One of his tools is the rooster, which one ancient text says "raises voice and calls men to prayer." Such Zoroastrian beliefs penetrated much of West Asia and India starting in the sixth century BC, as the empire's good roads and stability sparked a trade boom by linking the Indian subcontinent with the Mediterranean Sea for the first time since the days of ancient Ur, which itself underwent a modest renaissance. A Persian coffin found near its silted harbor contained a tiny seal with the image of a triumphant rooster.

The Persian prophet's view of life as a constant struggle between light and dark, good and evil, and truth and deception deeply influenced Judaism, Christianity, and Islam. Before the Persians came to Palestine, there was no Satan opposed to God, no hell to burn in, and no apocalypse to await. The only religious authorities said to have been present just after the birth of Jesus were not Jewish rabbis or Greek philosophers but the Zoroastrian priests called magi. Chickens are absent in the biblical Old Testament, but Christ mentions roosters and hens in the New Testament.

A couple of centuries after Cyrus's capture of Babylon, the bird had spread from Sudan to Spain, reached Kazakhstan in distant Central Asia, on the rim of Persian influence, and may even have braved the Atlantic with seafaring Phoenicians eager to exchange poultry for English tin. No longer solely an exotic gift, the chicken became enmeshed in the religious beliefs and practices of the ancient Western world. In Greece it became the sacred animal of a half-dozen gods and goddesses, and during Rome's heyday it predicted the outcome of battles. The rooster's crow marked the apostle Peter's betrayal of Jesus in Jerusalem on Good Friday, and followers of the cults of Mithras and Isis sacrificed it in temples from Egypt to Britain. By early medieval times, by papal decree, it pointed the wind's way on the churches of Christendom.

Islam gave it special rank as well. "When you listen to the crowing of the cock," the prophet Muhammad would tell his followers a millennium after the rise of the Persian Empire, "ask Allah for His favor as it sees Angels." According to some Islamic traditions, Muhammad saw an enormous and indescribably beautiful rooster standing on the foundation of the seventh and lowest level of earth with its head in the heavens, proclaiming the glory of God.

Amid the flocks of geese and doves, ibis and partridge, crows and vultures found all over the Middle East and Europe, the chicken became the premier sacred bird of awakening, courage, and resurrection. This triumph over so many competitors took place within a few short centuries. One scholar thinks that the very shape of the chicken reminded the ancients of an oil lamp—the common source of artificial light in the ancient world—with its spout resembling a protruding beak and its handle the upright tail. Others point to the hen's productivity and the fierceness of the rooster as potent symbols of fertility and war. Its crow, of course, encouraged farmers to quit their beds and produce sustenance for their communities and revenues for the state. The bird's origin in the mysterious and faraway East and its long tradition as a royal bird also may have set it apart from more prosaic farmyard animals.

Yet even in distant China, the chicken was associated with the sun and the conquest of light over darkness. This may reflect Zoroastrian influence that stretched all the way to the Middle Kingdom, since one empress worshipped a god identified as Ahura Mazda in the early centuries AD, when Persians traded as far east as the Pacific coast. A Chinese legend from the third century AD claims that the bird descended from the Vermilion Lord, a human who changed himself into a chicken. Daoist priests in this period sacrificed chickens to consecrate a new temple, ward off evil spirits threatening the imperial household, and drive away epidemics. By holding a rooster to his mouth—a peculiar practice still common among cockfighters—a priest could breathe out unwanted demons. A *jiren*, or chicken officer, was responsible for the sacrificial birds, and maintaining a flock of different colors to meet the needs of various rituals.

In that time, even the voice of the rooster was considered regal. When Chinese rulers wanted to announce a general amnesty, the imperial guard erected a giant cock with a head of pure gold in front of the palace on a post under a richly decorated pavilion. Thousands of people would vie to take a bit of earth from around the post to gain luck. Even today, one of the most prominent film awards given in China is a statue of a golden rooster. The Chinese ideogram for rooster sounds like the one for good omen, and the bird is one of the twelve zodiac signs. People born under its sign are considered to be keenly observant truth tellers. In early medieval Korea, chickens were raised in the royal court, and a white rooster is said to have heralded the birth of the founder of a clan and dynasty.

In Japan, by the seventh century AD, white chickens sacred to the great Japanese Shinto goddess of the sun, Amaterasu-ōmikami, roamed temple grounds. They were the sole animals capable of drawing her out of the cave where she hid. The western Chinese minority group called the Miao still tell the story of the world's early days, when the six suns refused to come out because they feared an archer would shoot them. No one knew what to do. Then a rooster appeared, quite literally saving the day. The small bird coaxed the sun to shine, simply by crowing. A team of Japanese researchers recently determined that this crowing originates from a sensitive circadian clock in the rooster that registers light before humans do.

From Germanic graves to Japanese shrines, the chicken emerged at the start of our common era as a symbol of light, truth, and resurrection across dozens of religious traditions spanning Asia and Europe. Tibetan Buddhists shunned the bird as a symbol of greed and lust, perhaps because the animal until recently was impractical to rear on the cold Tibetan plateau. Across most of the Old World, however, the chicken's growing spiritual role demonstrated not just the fowl's remarkable utility on the farm but its capacity to reflect our changing beliefs. Like the Yezidi, the bird adapted as empires and religions rose and fell. Gods, creeds, and dogmas appeared, vanished, and transformed, but the chicken became a constant and essential part of our worship.

3.

The Healing Clutch

You see this egg? That's what enables us to overturn all the schools of theology and all the temples of the earth!

—Denis Diderot, *D'Alembert's Dream*

As the poison numbed Socrates's feet, legs, and then groin, the Western world's most famous philosopher turned to his friend Crito and said, "We owe a cock to Asclepius," referring to the Greek god of healing and medicine. "Pay the debt. Don't forget." His weeping friend agreed, and, moments later, when the potion reached his heart, Socrates expired. The last words of a condemned revolutionary thinker, who was praised by his famous student Plato as "the best and wisest and most righteous man," were about a chicken.

In ancient Greece, sacrificing a cock to Asclepius was a common practice of a sick person who wanted to recover his health or celebrate its return. The German philosopher Friedrich Nietzsche was sure that Socrates was making an ironic comment on life as a terminal disease. Others say he was expressing his pious belief that the sacrifice would ensure his immortality, since Asclepius was capable of raising the dead. The classicist Eva Keuls argues that the optimistic and irreverent philosopher was telling a dirty joke to cheer up his grieving companions. Before he died, she contends, Socrates lifted his cloak and exposed his erection, which resulted either from the poison or the touch of the attendant who felt his body to check the drugs' progress—or both. He was making a pun on the Greek word for becoming cold, which can also mean rigid or enlivened, while referring to a bird associated with an insatiable sexual appetite as well as healing.

In Socrates's day in the late fifth century BC, the chicken was called the Persian bird. "Persian cock! Good Herakles! How on earth did he manage to get here without a camel?" says a character in *The Birds* by Aristophanes. Three centuries earlier, when Homer likely composed his famous sagas, the canny hero Odysseus didn't encounter chickens on his long and treacherous voyage, which took him from Turkey to Egypt, to various Mediterranean islands, and finally to his island home off the Greek mainland. By 620 BC, the first images of the bird appear on Greek vases. A detailed and lifelike terra-cotta rooster from that era was found in Delphi, near the famous oracle and sanctuary dedicated to Apollo, god of the sun, light, and truth.

The chicken in Greece was a powerful symbol of healing and resurrection. Aesop's goose laid a golden egg, but a lesser-known tale attributed to him, "The Cock and the Jewel," may be the oldest surviving story starring a chicken. In it, a rooster comes across a precious gem and recognizes its worth but realizes it is of little use to him. "Give me a single grain of corn before all the jewels in the world," the wise creature concludes. Some ancient Greeks, like Persians and Indians, considered the bird too sacred for killing. As in Babylonia, it was associated with solar and lunar deities. "Nourish a cock, but sacrifice it not; for it is sacred to the sun and moon," advised the mystic and mathematician Pythagoras in the sixth century BC. The bird also was associated with Persephone, the goddess of renewal who spent half her time in the underworld, bringing spring with her when she returned.

These beliefs almost certainly reflect older views across the Aegean Sea in Asia. At the time of Socrates's death in 399 BC, the Persian Empire stretched from Pakistan to the Hellespont, separating Asia from

Europe. Though Greek propaganda paints the Persians as tyrannical degenerates, new goods, plants, animals, ideas, religions, and inventions were welcomed. In Athens, Persian clothes, architecture, and food were fashionable. A deliciously juicy new fruit called the Persian apple was popular in the stands of Athens's marketplace, though the peach—its scientific name is still *persica*—actually originated in western China and worked its way east on Persian-controlled trade routes.

The chicken's association with the superpower to the east made it a perfect foil for Aristophanes, who enjoyed poking fun at contemporary philosophers like Socrates as well as the political authorities of his day. "It was he"—the rooster—"who was the first king and ruler of the Persians, well before all those Dariuses and the Megabuzes," says another character in the zany comedy, referring to the kings and priests of the neighboring empire. "That's why he's still called the *Persian bird* . . . he struts about like the Persian King!" Aristophanes's strutting rooster wore heavy armor and a high comb that resembled the crown of the Persian king and sported, according to stage directions, an "extra-long, extra-red phallus."

The outrageous costume poking fun at the neighboring imperium surely delighted the male audience in a time and place when a smaller penis was considered more beautiful and less barbaric than a large one. And the rooster's inevitable and insistent crow was an annoying reminder of its monarch's power to command. "As soon as he sings his morning erection, everyone else has to get up too and go off to work," complains one character. "The metal worker, the potter, the skin stretcher, skin puller, skin washer, whore, lyre and shield maker—all get up, still in the dark, put on their shoes, and off they go!"

Produced fifteen years before Socrates was sentenced to death for corrupting youth and spreading impiety, the fantastical play is itself heretical, recalling a time when "it wasn't the gods who were the kings and ruled over humans but you lot"—the birds. With the help of two Athenians, they build a city in the sky, reclaim their lost privileges, and lead a successful revolt against the Olympians. New laws forbid trapping, shooting, and eating birds. Guarding the walls of the new celestial acropolis was "Ares' Killer Kid," the fierce Persian bird favored by the god of war.

Aristophanes's satire about a power-drunk, bellicose, and sex-crazed bird reflects the multifaceted role that the chicken played in classical Greece, which adopted many of the older Near Eastern myths surrounding the creature. Because we have so few records from the Babylonian and Persian Empires, our first comprehensive view of the Western chicken comes from its early European beach-head. Roosters on Greek vases perch like those on earlier Babylonian seals, propped on columns in scenes of worship. They flank Athena, the goddess of wisdom, courage, and the peaceful arts, or are emblazoned on her armor. The helmet of the famous gold-and-ivory statue of Athena on the Acropolis was adorned with a cock. Roosters also show up in images of Hermes, the protector of gamblers and athletes.

Laying hens were common by Socrates's day, but pork and goat meat were far more available and popular than chicken. What set the bird apart from other animals, aside from its formidable fighting and sexual abilities, were its close ties to Asclepius, the half-human progeny of Apollo, the supreme god of light and healing. In the original Hippocratic oath, a doctor would swear by both Apollo and Asclepius, whose cult appears to have begun in an Apollo temple in Epidaurus, just across the Corinthian isthmus from Athens, about the time the chicken arrived in Greece by the seventh century BC. Small and slender vases from the seventh and sixth centuries BC picture a wriggling snake between two roosters. Called alabastrons, these little vessels held ointments and perfumes that also may have had medicinal value, since the snake was the other primary sacred animal associated with Asclepius.

By the time of Socrates, the god was a popular deity, and his temples evolved into the premier spas and healing centers of the Mediterranean world for the next eight centuries. Families would gather here to pray for or celebrate the recovery of a relative with sumptuous feasts that typically involved chicken sacrifice. There was a spacious Asklepion on the south side of Athens's acropolis, not far from the theater where many of Aristophanes's plays were performed.

At Epidaurus—often called the ancient Lourdes—the facility had 160 rooms for patients along with a sanctuary, and the one in Pergamon in Asia Minor was a luxurious complex for well-to-do clients boasting its own theaters and sports facilities. At the Asklepion on the island of Kos, where Hippocrates received his medical training, a third-century BC Greek poet records that two women sacrificed the rooster they had brought, cooked it, carved a drumstick for the priest, gave the sacred snake on the altar a morsel, and then packed up the leftovers to eat at home.

The apostle Paul's prohibition on Christians eating "idol food" was one factor in the decline of the Asklepion in the latter days of the Roman Empire. All that remains of this era, aside from the archaeological ruins and our words *hygiene* and *panacea*—derived from the names of Asclepius's daughters—is the wriggling snake coiled around his rod on the back of North American ambulances. But as late as the third century AD, initiates in the Eleusinian mysteries consecrated a cock to Asclepius or to Demeter, the goddess of the earth and the harvest, while abstaining from eating chicken during their rites.

For thousands of years, across many cultures, the bird served as a remarkable avian medicine chest. Its meat, bones, organs, feathers, comb, wattles, and eggs all figure frequently in ancient prescriptions. No matter what ailed—migraine, dysentery, insomnia, asthma, depression, constipation, severe burns, arthritis, or just a nagging cough—the chicken provided an all-purpose, twenty-four-hour, two-legged drugstore. Galen, the second-century AD Greek physician, prescribed dried rooster gizzard with juniper juice for bed-wetting. Others recommended chicken brains to make an infant's teeth grow or as an antidote to a snakebite. Cock dung, it was said, could cure an ulcerated lung. The eleventh-century Persian philosopher Avicenna served soup made from a young hen as a remedy for leprosy, and a French Renaissance recipe calls for squeezing a chicken and drinking the resulting juice to cure a fever. "Chicken offers so great an advantage to men in its use in medicine that there is almost no illness of the body, both internal and external, which does not draw its remedy from these birds," wrote a sixteenth-century Italian scientist.

In the ancient city of Milas on the Aegean coast of what is now Turkey, raw chicken egg whites are still used to remedy burns, as recommended from Hippocrates to twentieth-century American physicians. A 2010 study of the practice in Milas published in the *Journal of Emergency Nursing* found that parents who used egg whites on their children's burns exacerbated the chance of infection. It is better to put the burn area in cold water and then cover with sterile gauze. Chicken soup, however, has stood the test of time and modern science. The Roman writer Pliny called it very effective for treating dysentery. A 2000 study in *Chest* found that the soup—the researcher used the recipe passed down by his wife's grandmother—was loaded with ingredients that together produce a mild anti-inflammatory effect mitigating the upper respiratory symptoms common from a cold. Another study confirmed that chicken soup cleared up the stuffy noses and chests of volunteers more effectively than hot water. A third investigation found that it strengthens the cilia in the nose that halt potentially damaging bacteria and viruses. And a 2011 study by an Iowa physician determined that a group of people sick with a viral illness who ate chicken soup—even store-brought brands—recovered faster than those who did not.

Chicken meat contains cysteine, an amino acid that is related to the active ingredient in a drug used to treat bronchitis, which might offer a clue to the soup's long-standing fame. Some researchers suspect that the soup reins in the immune system, and by doing so slows the inflammation that is the body's response to the viral attack.

Other more esoteric prescriptions have also been borne out by research. Rooster combs, for example, can indeed ease arthritis and smooth wrinkles. The combs, it turns out, are a rich source of hyaluronan, a compound that can reduce inflammation and has been used on racehorses for several decades. The pharmaceutical company Pfizer now breeds White Leghorns with enormous red combs to harvest the substance for patients with creaky knees. Its rival Genzyme uses the compound in a gel designed to act like Botox, puffing out sagging, aged skin. And proteins derived from chicken bones have been shown to halt the pain of those with rheumatoid arthritis. But the bird's most vital behind-the-scenes role in human health is combating influenza, one of our most common and deadly scourges.

As a beeping truck backs into the loading dock in a cold drizzle, Peter Schu asks me to guess how many eggs the vehicle carries. About half the size of an American tractor trailer, it seems small but I guess high. "Fifty thousand," I venture. I know by his smile that I'm way off. "One hundred and eighty thousand," says the director of this GlaxoSmith-Kline pharmaceutical plant.

Each day at precisely 6 a.m. and at midnight, a black metal gate swings open to let an unmarked truck like this one rumble into a factory complex in the city of Dresden, a hundred miles south of the German capital of Berlin. The cargo is from dozens of farms scattered around the German and Dutch countrysides that produce a total of 360,000 eggs daily. Their locations are secret.

The eggs are the test tubes that produce the vaccine that inoculates us from influenza. Hippocrates described flu symptoms nearly twenty-five hundred years ago, and pandemics were first recorded in the sixteenth century. In a normal year, the flu virus sickens millions and kills between 250,000 and 500,000 people. In the global pandemic of 1918, an estimated 50 million people died. About one in three humans fell ill, and five hundred thousand died in the United States alone. That pandemic may have begun in chickens, moved to pigs, and then spread to humans. Influenza appears to be a price humans have paid for domesticating animals, and the illness continues to evolve and spread among our species. A new pandemic is always a global threat.

Ironically, the chicken today provides us with our main source of the world's vaccine supply for a disease that continues to cycle between birds and humans. Combating this fast-evolving virus requires a concerted international effort by teams of doctors, epidemiologists, microbiologists, and public health officials. Each year, scientists around the world collect samples and make educated guesses as to which strains are likely to spread during the coming winter. The vaccine uses tiny doses of the chosen inactivated strains to trick the body's immune system into producing antibodies, which can then fight off the infection when encountered during flu season. The selected viruses are cultured and sent to manufacturing plants like this one in Dresden.

Schu oversees the seven hundred workers who pump out 60 million doses annually and distribute them to seventy countries before the flu season takes hold. To do that, he needs a steady supply of fertilized chicken eggs. Each one must be kept scrupulously clean from the moment it is laid, and rigorous security precautions ensure a steady and sterile supply from farm to factory. Even he is forbidden from visiting the laying facilities. The farmers pledge to keep their work secret even from their neighbors, and no posted signs hint at their unusual product. Bacteria and virus brought by outsiders could delay or halt the clockwork production, endangering output and creating a potentially disastrous shortfall in global vaccine distribution. When a dangerous avian flu spread to a poultry operation close to one of the specialized farms in 2008, all the chickens in a ten-kilometer radius were killed and Schu had to struggle to find eggs from other farms to make up for the shortfall. "That's why you don't put all your eggs in one basket," he adds wryly.

As the driver shuts off the engine and climbs out of the cab for a smoke, Schu takes me into a small white-painted locker room in the main factory building. The vaccines are made in a sleek new glass-and-steel structure that takes in more than a third of a million eggs daily, each nearly identical in size and shape, laid almost exactly nine days earlier. Each has already been checked to ensure a regular shape and weight—only eggs that are fifty-four to sixty-two grams are acceptable—and that an embryo is developing.

Schu demonstrates how to wash hands thoroughly and slips his lean frame into a full white zippered jumpsuit and powder-blue booties to cover his bacteria-laden street shoes. Then he dons enormous clear goggles. This must be done without any article touching the floor contaminated by street shoes. It takes me twice as long to get into the awkward gear. Suited up, we sit on a bench dividing the room in half, swing our legs over, and head for the only door on the other side. He pauses. "This is the last step before we enter areas with the active virus," he says in a somber tone.

We enter a corridor and take an elevator to the third floor, where a long glassed-in hallway overlooks the loading dock below, which is more akin to the air lock in a spacecraft, providing a seal between the outside world and the eggs stacked inside the truck. At the end of the hallway, workers shrouded in the same coverings that we wear trundle spiffy stainless-steel versions of movable cafeteria cabinets, each containing hundreds of eggs nestled in plastic trays. In a small anteroom off the hallway, we put on yet another layer of gloves and booties.

When Schu opens the next door, we step into a small rectangular room with a wall of glass on one side, and a large door like that of an enormous walk-in freezer on the other. The place looks like a small, efficient institutional kitchen. Half a dozen women are busy moving trays filled with eggs as Schu opens the giant door. A blast of warm and humid air dissolves the Dresden chill still in my bones. Inside are racks and racks of eggs. The new arrivals are stacked in this enormous incubator overnight so that they can adjust after their bumpy journey.

Once the eggs have settled, the white-suited workers load the egg trays onto two conveyor belts that feed into a large room behind the glass. Resembling a modern German brewery with its stainless-steel tank for the live virus and assembly line machinery, it is off-limits to all but a small number of workers who oversee the operation of a steel needle filled with the virus solution that precisely and delicately punctures each egg. Then the ovum is incubated for up to seventy-two hours so that the injected virus can infect the embryo and multiply in the nutrient-rich but sterile environment. The fluid in the egg is harvested—chemicals render the virus inactive—purified using high-speed centrifuges, separated into individual doses, squirted into syringes, shrink-wrapped, packaged, and distributed around the world. The flu shot that you get at the local drugstore can be traced back to a plant like this one.

The U.S. Army in World War II developed this technique amid fears that a reprise of the 1918 pandemic might paralyze the war effort against German and Japanese forces. American soldiers received the first flu shots in 1945 as Dresden was firebombed by British and American planes. Millions of people have since avoided illness and death as a result of annual shots. Since each dose requires an average of three eggs, however, the process remains complicated and expensive. One egg infected with a pathogen can render useless an entire batch of vaccine.

Until recently, this has been the only practical way to produce large amounts of the vital medicine, but in late 2012, the U.S. Food and Drug Administration (FDA) approved an eggless vaccine that instead grows the virus in mammalian cells. Early the following year, the FDA also gave the green light for a simpler, less expensive process that doesn't even use live influenza virus. Instead a genetically modified virus infects the cells of insects to produce a protein that can trigger your immune system to make antibodies just like the egg-based version. "There will be a shift to an egg-free process," Schu concedes as we take off our protective gear back in the factory locker room. "But the new approaches still have a number of hurdles to overcome."

Even after the egg is no longer needed for influenza vaccines, it will remain the model animal research organism. Deceptively simple on the outside, this perfectly contained system is more complex than any pressurized spacecraft. Under its smooth exterior shell is an inner and outer membrane and an air cell that balloons at the flatter end to allow the system to expand and contract. Under this are two thin layers of slightly denser albumen—the egg white—encasing the clear liquid that takes up the bulk of the interior. Two strands of tissue connect the yolk, enveloped in its own membrane, to either end of the egg's interior. In a fertilized egg, the embryo feeds off the yolk, deposits waste in a sack, and takes in oxygen through the semipermeable shell that simultaneously keeps out any intruding bacteria or virus. Perfectly protected, the embryo has no need for an immune system and antibodies until three days before it hatches.

Aristotle studied chickens and in the process hatched embryology. He observed the mating habits of roosters kept apart from hens in temples—possibly in the Athenian Asklepion that had chicken flocks for sacrifice. He also cut little holes into fertilized eggs so that he could record the growth of the embryo over the three weeks it takes for an egg to hatch. The observations allowed him to dismiss the notion, held by many scientists well into the nineteenth century, that the embryo was an animal in miniature that simply got bigger. Instead, the Greek thinker proposed that a fetus develops in clear stages. These innovative chicken-egg experiments led him to examine other species, including humans, until he concluded "all animals whatsoever, whether they fly or swim or walk upon dry land, whether they bring forth their young alive or in the egg, develop in the same way."

The tiny universe of the chicken egg remains the miniature laboratory of choice for embryologists. William Harvey in seventeenth-century London used the egg to understand how blood flows and to trace what became known as the nervous system. Marcello Malpighi at Bologna used the newfangled microscope to describe capillary vessels and other key elements of developing anatomy in chickens. Three centuries later, in 1931, scientists cultured viruses in fertilized chicken eggs, paving the way for the first cost-effective vaccines to combat mumps, chicken pox, smallpox, yellow fever, typhus, and even Rocky Mountain spotted fever—and, eventually, influenza.

By the 1950s, cancer researchers used chicken eggs to understand tumor development. By inserting cancer cells into a fertilized egg, they examined how the disease feeds and spreads. One of a new generation of scientists, Andries Zijlstra at Vanderbilt University Medical Center in Nashville, has found an innovative way to observe tumors grown inside an egg. "The trick has been to keep an animal alive while watching a tumor grow," he tells me when I visit his lab, a short drive from the bars of the country music capital.

The son of a Dutch farmer, Zijlstra pioneered a way to decant a young embryo into a plastic petri dish without disturbing its growth by maintaining proper temperature and humidity. After injecting tumor cells into its blood vessels, he then takes a photo every fifteen minutes as the cancer cells proliferate, and watches how cells respond to cancer. In one corner of his lab, Zijlstra pulls out a rectangular tray with a couple dozen shallow depressions, inside of which are bits of what look like orange Jell-O that has not quite hardened. "These need to cook for a few more days," he says of the embryos, meaning that they require further incubation. His lab goes through twenty thousand eggs a year, produced in near-sterile conditions like those used in Dresden. A single one costs three dollars, about the average price of a dozen in a store. "With the chick embryo, you can actually see what is going on, unlike with a mouse or even a zebra fish," he tells me as he slides the tray back into its warm and humid home. "It is a fully intact biological system, not one that has been chopped up into pieces."

On October 30, 1878, a package containing the heart of a young rooster arrived at the Pasteur Institute in Paris. The bird had died in Toulouse after being injected with a deadly disease then sweeping French flocks. At that time, the only vaccines available were for illnesses like smallpox and cowpox. The unaltered virus usually stimulated antibodies and protected the patient, but it could also kill.

At first Louis Pasteur, a world-famous and busy scientist in his midfifties, paid little notice to the delivery. But his young assistant Charles Chamberland bought two live chickens in the Paris market and injected them with the virus drawn from the dead cock's heart. When he returned to the institute the next morning, they were both dead. Chamberland wanted to culture the microbe, but his efforts using a mixture of yeast and water failed. Just after New Year's Day 1879, he and Pasteur hit on that all-purpose remedy: chicken soup, or, more precisely, chicken broth. "We now have a culture medium for these little organisms," Pasteur noted with excitement. The bacillus survived in the broth, since live chickens inoculated with it quickly expired.

They made no further progress until summer. Like most well-to-do Parisians, Chamberland was eager to begin his August vacation. Delaying plans to inoculate several hens with the "little organisms," he absentmindedly left the broth sitting out near the chicken cages while away. When he finally got around to the task several weeks later, the birds sickened but recovered. Intrigued, Pasteur ordered another set of injections. "Chance only favors the prepared mind," Pasteur once said. When the birds did not die, and ones bought that day in the local market did, he realized that the original fowl had developed a resistance to the disease after being exposed to small amounts of the microbe drifting in the lab air during the summer break.

That revelation led to an intensive effort to manipulate the virus so that it would produce antibodies without turning lethal. The scientists experimented with increasing acidity and decreasing temperature in the infected broth and observed the resulting effect on the health of chickens procured from the markets. Exposing the bacillus to oxygen weakened its impact. By January of 1880, Pasteur was able to inoculate chickens—at one stage as many as eighty in cages at the lab—with the altered virus, which gave them immunity without making them ill. "Through certain changes in modes of culture development," he triumphantly informed the Academy of Sciences the next month, "we can reduce the virulence of the infectious microbe."

This was good news for the French poultry industry, faced with devastating losses from what was then called chicken cholera. In the end, however, inoculation proved more expensive and less efficient than isolating and slaughtering infected birds. More important, the development of the first artificial vaccine marked the start of a revolution in human medicine that would help researchers combat diseases that annually killed millions of people.

A dozen years later and halfway around the world, another set of chickens provided a critical insight into human diet and disease. A Dutch doctor in Indonesia, Christiaan Eijkman, fretted over the frequent occurrence of beriberi, a painful illness that swells legs and can lead to heart failure, and like Pasteur, he made a serendipitous find. Budget cuts at his army hospital forced him to change the way he fed his flock of chickens kept for eggs and the occasional chicken soup. He bought cheaper gray unpolished rice for some, while others received the white rice leftovers from the dining hall. Over time he noticed that the former stayed healthy while the latter came down with the disease, discovering the importance of vitamin B, for which Eijkman and another scientist eventually shared a Nobel Prize.

Chicken experiments paved the way to curing illness, but were also used by the originators of the eugenics movement in the early twentieth century who sought to rid humanity of what they saw as degenerate traits. In 1910, the director of New York's Cold Spring Harbor Laboratory, Charles Davenport, hired a Missouri chicken breeder named Harry Laughlin, who used his influential position as managing director at the eugenics office to convince Congress to limit immigration of Eastern Europeans. Laughlin saw chicken breeding as a template for improving the human species through scientific selection. He helped push through laws in eighteen states imposing compulsory sterilization on a range of physically disabled and indigent persons. In 1933, lawmakers in Nazi Germany decreed a similar sterilization law based on Laughlin's legal language. Hundreds of thousands of people, in both Germany and the United States, were forcibly sterilized. Both Davenport and Laughlin died before the end of World War II, and eugenics was thoroughly discredited by the policies of the Nazis.

Another eugenicist in this era argued that cells had extraordinary longevity based on his chicken experiments. The French biologist Alexis Carrel, who pioneered the technique of suturing arteries, earned a Nobel Prize and a position at the Rockefeller Institute for Medical Research in New York City. On January 17, 1912, he sliced a piece of heart from an eighteen-day-old rooster embryo and kept it alive in a culture of clotted chicken blood as it divided into new cells. This experiment electrified the American public, and the *New York Times* reported annually on its growth and health. Carrel was a member of a scientific committee that proposed sterilizing up to 8 million Americans considered unfit to procreate. He returned to France at the outbreak of World War II to work for the Germans' puppet government based at Vichy and died of heart failure in 1944. The famous still-beating immortal chicken heart in New York survived two years longer before it was discarded when his assistants moved on to other laboratories. Only later did researchers realize that the chicken blood regularly fed to the culture gave the heart cells the illusion of immortality.

A century after Carrel, biologists seek to transform the bird into a miniature drug factory. Human proteins make antibodies to counter illness, but producing these is expensive and complicated. Egg white offers a tantalizingly cheap and easy way to mass-produce them, as the eggs in Dresden already do by serving as tiny bioreactors. Unlike other species such as goats or hamsters, chickens make proteins in a remarkably similar way to humans. By inserting genes from other species, including humans, into what is called a transgenic bird, researchers hope to manufacture protein-based drugs for ovarian cancer, AIDS, arthritis, and a host of other illnesses at a fraction of today's cost.

At Edinburgh's Roslin Institute, famous for cloning Dolly the sheep, biologists inject a human antibody into the embryos of fertilized eggs. The embryo then is transferred to a host egg, and some of the hatched birds produce progeny that retain the alien DNA. Other scientists in the United States are working on altering rooster sperm so that the antibodies can become part of its genome and then be passed on to a fertilized ovum. These approaches might pave the way for Dresden-like factories churning out cheap human-protein-based drugs with genetically modified eggs. Twenty-four centuries after Socrates asked his friend to sacrifice a cock to Asclepius, the era of the transgenic chicken is quietly dawning, coming soon to your bathroom medicine cabinet.

4.

Essential Gear

Next to the Dog, the Fowl has been the most constant
attendant upon Man in his migrations and his occupation of
strange lands.

> —Edmund Saul Dixon, *Ornamental and Domestic
> Poultry: Their History, and Management*

The chicken's greatest journey was from west to east, across the
Pacific Ocean in what was also humanity's greatest feat of exploration
prior to the sixteenth century. The Pacific was long the single biggest
obstacle in the spread of our species out of Africa. We did not need
cultivated plants, domesticated animals, or knowledge of the stars to
settle most of the planet. Fifty thousand years ago we already had left
Africa, crossed Asia, and paddled to Australia. But thirty thousand
years ago, our migration halted on the Solomon Islands scattered just
to the east of New Guinea, on the threshold of the deepest and widest
ocean. Humans then moved around the Pacific, pivoting north to
cross Siberia into Alaska and fill out every corner of the Americas all
the way to Tierra del Fuego by at least thirteen thousand years ago,
and likely much earlier.

For another ten millennia, this vast Pacific region that took up nearly one-third of the planet remained beyond our grasp. We needed seaworthy vessels and advanced navigational techniques as well as carefully selected crops and hardy but compact animals that could survive a long voyage in cramped quarters. Not until after AD 1200 did the double-hulled canoes of Polynesians reach the remote Hawaiian archipelago and distant Easter Island. And chickens were an essential element of the gear that people took along.

"How shall we account for this Nation spreading itself so far over this Vast ocean?" wondered Captain James Cook, who commanded the first European ship to reach Hawaii half a millennium after its settlement. Well into the twentieth century, many Westerners claimed that Polynesians must be the remnant population of a sunken continent, trapped on the remaining bits that did not submerge. Others suspected that a hidden southern land just below the islands provided an easy jumping-off point for the simple seafarers. Few imagined that this people could have accomplished such an astonishing feat of colonization without modern compasses, sextants, and large ships.

Cook saw their skill firsthand. At the insistence of the naturalist Joseph Banks, a correspondent with Linnaeus, he took on board the HMS *Endeavour* an artist, priest, and politician named Tupaia while cruising near Tahiti. Lacking all charts and instruments, Tupaia was familiar with well over a hundred islands in a two-thousand-mile radius and knew his location as they moved across the South Pacific to New Zealand. "These people sail in those seas from Island to Island for several hundred Leagues, the Sun serving them for a compass by day and the Moon and Stars by night," Cook marveled while still on Tahiti. He believed that further investigation would reveal the origin of the audacious venture. "When this comes to be prov'd we Shall be no longer at a loss to know how the Islands lying in those Seas came to be people'd . . . so we may trace them from Island to Island quite to the East Indies."

Aboard the *Endeavour*, chickens were housed in a hutch in front of the ship's wheel and were an important source of meat and eggs during Cook's lengthy voyages. For the Polynesians, they were something more. Upon arriving at one island in 1769, Banks recorded that at their reception a distinguished old man "immediately ordered a cock and hen to be brought which were presented to Captain Cook and me, we accepted of the present." Easter Islanders made a similar offering to the first Europeans to arrive on their shores in 1722 under less friendly circumstances. Within minutes of landing, Dutch sailors shot and killed a dozen unarmed natives for making what they perceived as threatening gestures. The fearful chief ordered chickens given to the expedition leader to prevent further bloodshed. That evening, the islanders brought both live and roasted birds to the ship in order to appease the dangerous foreigners.

"Chickens played an important role in native life, and the remains of the dwellings made for them are much more imposing than those for human beings," the British archaeologist Katherine Routledge, who surveyed Easter Island's ruins on the eve of World War I, wrote in her 1919 book, *The Mystery of Easter Island*. She recalled that an honored native frequently was asked to perform a ritual to increase the fertility of people's fowl. As Europe teetered on the brink of its Great War, Routledge found herself trying to stop bloodshed between natives and Chilean sheep ranchers. The leader of the native rebellion, a shaman priestess named Angata, sent her two chickens to gain her backing in the uprising.

By Routledge's day, Polynesians on one of the world's most remote inhabited islands were a remnant of the original population, which had been decimated by disease or deported to Chilean mines. Extreme isolation, a small population, a windswept landscape, and eerie sixty-foot-high statues make Easter Island a Rorschach test for the Western world. The Manhattan-sized island is today the poster child of ecological disaster brought on by human population growth and material greed, a cautionary tale for a planet in crisis.

"Humankind's covetousness is boundless. Its selfishness appears to be genetically inborn," a New Zealand and a British archaeologist wrote in *Easter Island, Earth Island*. "But in a limited ecosystem, selfishness leads to increasing population imbalance, population crash, and ultimately extinction." Author Jared Diamond echoed this grim warning in 1995. "Easter Island is Earth writ small." The islanders cut down their forests, ate plants and animals to extinction, and let their society collapse into tribal warfare and cannibalism. "For domestic animals," he wrote, "they had only chickens."

Scientists, writers, photographers, and tourists flock to see and study the mute and massive statues, but they usually ignore the chicken coops. When colonists weren't carving giant statues, they seemed to have created coops as part of a sophisticated system of growing crops and caring for fowl. Hundreds of these structures are still scattered across the island, neatly built dry-stone cairns, each with a small entrance secured by a stone door. They were often part of or adjacent to hundreds of walled gardens—some shaped like large canoes—that sheltered crops from the blustery Pacific winds. The stout walls protected the birds from their only predator on the island, the rat, as well as bad weather, and the manure may have provided critical fertilizer for the nearby fields.

There is no definitive proof that these coops predate Europeans' time there. Terry Hunt, an archaeologist at the University of Oregon who has conducted years of digs on the island, thinks they are of ancient Polynesian origin. The coops served as stone quarries to build the nineteenth-century ranch walls that still crisscross the island, and Hunt thinks it unlikely that hundreds of these small buildings could have been constructed in the previous century, when the native population plummeted. The bird had religious, political, and agricultural significance for islanders, and the destruction of their coops to fence imported sheep may have led to the rebellion that Routledge attempted to quiet.

In his 1997 book, *Guns, Germs, and Steel*, Diamond acknowledges that the islanders practiced "intensive poultry farming," but he is skeptical this made a difference. Even with dogs, pigs, and chickens, Polynesians on other islands could enjoy only "occasional meals" with protein-rich meat. Crops were the key to survival. This view may seriously underestimate the dietary role of chickens and the impact that their manure may have had in fertilizing fields, though researchers have yet to gather the necessary data on the extent of poultry farming on the island before Europeans arrived.

Easter Island was never an easy place on which to survive. "No Nation will ever contend for the honour of the discovery of Easter Island as there is hardly an Island in this sea which affords less refreshments, and conveniences for Shiping than it does," wrote James Cook when he arrived in March 1774, a half century after the Dutch. He estimated that there were six or seven hundred inhabitants, but didn't know that this population was about a quarter of what it had been when the first Europeans arrived. Hunt argues that European disease was the prime mover behind the rapid population decline, not native greed or folly.

The island's primeval forest had vanished prior to the Dutch visit. Rats, which gnawed palm-seed shells excavated by Hunt, were likely major culprits in the destruction of the forest. During Cook's visit, a member of his landing party observed "ratts which I believe they eat as I saw a man with some in his hand which he seem'd unwilling to part with." This has been interpreted as a sign of desperation and hunger, but Polynesians across the Pacific commonly ate the rodent. Cook's men also noted staples found on other islands, such as plantains, sugarcane, gourds, taro, and sweet potatoes—"the best of the sort I ever tasted." There were also "Cocks and Hens like ours which are small and but few of them." Entranced by the toppled statues, the British, like so many archaeologists who followed after them, may not have bothered to look inside the stone coops.

The discovery of what scientists claimed was a pre-Columbian chicken bone on the western coast of South America called for a major rewrite of human history. With the possible exception of the chicken bone lodged in her throat that sent Elizabeth Taylor to a Virginia hospital in 1978, no single piece of poultry cartilage has ever spilled so much newspaper ink. "Why did the chicken cross the Pacific Ocean?" asked the *New York Times* on June 5, 2007. "To get to the other side, in South America. How? By Polynesian canoes, which apparently arrived at least 100 years before Europeans settled the continent."

The news was heralded as scientific proof that East and not West began the exchange between the Old and New Worlds after being largely separate for more than ten thousand years. Polynesians trading poultry on a Chilean shore bridged the divide rather than Spanish sailors splashing onto a Bahamian beach.

The chicken remains were found at El Arenal—Spanish for "sandy place"—on the barren Arauco Peninsula that extends into the Pacific Ocean 250 miles south of the Chilean city of Valparaiso. Archaeologists working at the site found an unremarkable village inhabited for seven centuries until AD 1400. The inhabitants had left behind the usual assortment of pots, jars, plates, and bowls that they used to cook and eat shellfish, quinoa, frogs, ducks, corn, fox, and guanaco, a cousin to the llama. Among these remains were eighty-eight chicken bones.

According to traditional history, a pre-Columbian chicken feast in the Americas was impossible. By the time the Bering Strait opened, effectively sealing the New World from the Old, *Gallus gallus* was still a skittish wild bird in South Asia jungles. If there were truly pre-Columbian chickens, then New World and Old World humans had met sometime after the end of the Ice Age and before Columbus. Until 2007, no archaeologist had yet identified an actual bone, though tantalizing hints abounded of the chicken's presence when Europeans arrived in the Americas.

Shipwrecked Japanese sailors and medieval Irish monks are among the potential importers. The chicken arrived in Sweden by AD 500, and Icelandic breeders claim that one still-popular variety arrived with tenth-century AD Vikings. A couple of chicken bones dating to the late thirteenth century have turned up in Iceland, but there is no sign of the animal at the settlement the Vikings built on the Atlantic coast of Newfoundland. Nor is there evidence that native North Americans adopted the bird before Columbus landed. The first documented chickens in the New World arrived in 1493, when two hundred hens from the Canary Islands arrived with Columbus on his second voyage to Hispaniola—today's Haiti and the Dominican Republic—to create the first known New World poultry operation. A famine struck shortly after, and the island-bound birds likely were either killed for their meat or succumbed to predators and disease.

The Spanish conquistador Hernán Cortés and his troops arrived in Mexico in 1519. While dismantling the Aztec Empire, he told Spain's King Charles V that the natives roasted chickens as large as peacocks, sold them in markets, and fed them to the wild animals in Montezuma's private zoo in the center of what is today Mexico City. That zoo, built of marble and tiles of jasper, included a luxurious apartment for the Aztec king. Eagles and kestrels along with wildcats were fed on the fowl, Cortés reported. But he likely was referring to one of the New World's only domesticated birds, the turkey.

Cortés noted a street in the marketplace of the great city of Temixitan that sold "every sort of bird, such as chickens, partridges, quails, wild ducks" and a dozen other sorts. Since he doesn't mention the turkey, which surely was an important commodity, he likely classed it as chicken. Ancient Mesopotamians may have dubbed the chicken a "black francolin from India," and Cortés may have similarly named the turkey after its closest European look-alike. The Nahuatl-speaking Aztecs in central Mexico called Spain "the land of the chickens" since the bird came with the foreigners, a sign that they saw the two species as distinct and the chicken as alien. Since no chicken remains have been found in Mexico that predate Cortés, the bird likely came with the conquerors.

The evidence for chickens in South America before 1500 is ambiguous. When a Portuguese navigator that year bumped by accident into South America, about five hundred miles north of Rio de Janeiro, he reported many birds but no chickens. Shown a hen, some locals appeared terrified. Two decades later, an Italian adventurer named Antonio Pigafetta, who was traveling with Ferdinand Magellan on the world's first circumnavigation, landed on the Brazilian coast and traded one fishhook for a half-dozen chickens. Pigafetta adds that for a single king of diamonds from his deck of cards, "they gave me six fowls and thought that they had even cheated me."

In 1527, a Spanish expedition anchored off Santa Catarina— nearly seven hundred miles south of Rio—when its crew was struck with disease. The captain sent a man more than one hundred miles into the Brazilian interior to obtain chickens and other food for the ill in exchange for fishhooks, knives, and mirrors. Whether the native chickens reported by these early European visitors were really chickens is hard to discern. The South American jungle is home to the curassow, a bird that resembles a large chicken and was at least partially domesticated. In 1848, as the biologist Alfred Russell Wallace explored the Amazon, he watched curassows come and go at will around Indian villages, since "being brought up from the nest, or even sometimes from the egg, there was little danger of their escaping to the forest." In Venezuela, a gamecock breeder told me that curassows sometimes were bred with chickens to produce larger and tougher hybrids. Given the evolutionary distance between the two species, scientists doubt this is biologically possible.

One intriguing report on the existence of Amazonian chickens comes from a sixteenth-century German conquistador, Nikolaus Federmann, who explored the Orinoco basin to the north in 1530 on an unauthorized expedition to find gold. In a book published a quarter century later, he claimed that he heard a rooster crow in the Venezuelan jungle. Locals explained that the bird came from the Southern Ocean in a big house. Historians interpret this to mean that the bird made its way through the jungle from Portuguese ships on the Atlantic coast, but it is possible that Federmann's informants were referring to Polynesian canoes on the Pacific shore.

Just two years earlier, Francisco Pizarro arrived in the Andes Mountains on the western edge of South America. A half century later, a veteran of the campaign against the Incan Empire recalled encountering "a few white chickens of Castille" in Peru. No definitive archaeological evidence of pre-Columbian chicken has been found there, however, despite a century of excavations. But the very name of the overthrown emperor, Atahuallpa, is said by some scholars to be related to the Quechuan word for chicken. A museum in Lima displays a two-thousand-year-old terra-cotta vessel that looks very much like a rooster, complete with comb and wattles and upright tail. Art historians can't say for sure if it represents a bird that lived there at the time or was the fantasy creation of a local potter.

What is certain is that the chicken, if it was indeed a new arrival, adapted remarkably quickly to the continent. One Jesuit in the 1580s wrote that Indians avoided sheep and cattle, but embraced dogs and chickens. Men hunted with the dogs while the chickens were treated with exquisite care. "Women carry them on their backs, and raise them as children," he asserted, adding that the birds "were infinitely larger" than those in Portugal, a possible confusion with the curassow. Chickens soon were widespread and portable enough that they were often used as tax revenue. Some Brazilian natives were required to pay the Portuguese in the form of chickens, while hens and eggs were a regular form of tribute in Spanish-controlled areas.

By the close of the sixteenth century, eggs brought from the lowlands were used as currency in the high mountains and plateaus of Bolivia, Peru, and Mexico. Chickens were arriving in South and Central America from Europe, Africa, and Asia. Bantu slaves from southern Africa brought their varieties to Brazil as early as 1575. This mixing of genes poses a challenge to geneticists attempting to untangle the origins of the bird in the New World, which is why the discovery and dating of the El Arenal bones caught worldwide attention.

Those bones had gathered dust for more than two years in the Santiago house of the excavation director until a New Zealand archaeologist named Lisa Matisoo-Smith caught wind of the find while at a meeting in Chile. In January 2006, a Chilean colleague handed her a plastic Baggie containing one of the bones at the Santiago airport. Back in her lab at the University of Auckland, Matisoo-Smith turned the sample over to her PhD student Alice Storey, who broke the bone into three pieces. One piece went to an institute in Wellington for independent radiocarbon dating, which showed the chicken was alive and clucking sometime between AD 1304 and 1424. The second piece of bone went to Massey University in Auckland for DNA extraction, while Storey kept the third to do the same in her home lab. Both results then were sequenced at Massey.

No one at that time had any idea what genetic sequences were typical of Pacific chickens, so the team collected thirty-seven other chicken bones from Polynesia dating from 1000 BC to AD 1500 to test and compare their DNA. They also plucked feathers from Araucana chickens in Chile, odd birds with ear tufts and no tail feathers that lay light-blue and light-green eggs. Some breeders think their uniqueness reflects remnants of a pre-Columbian population. About one-third of the old bones from Polynesia yielded DNA sequences, including those excavated on Easter Island, two thousand miles west of the continent's coast and the likely stepping-stone to any South American contact by Polynesians.

The El Arenal bone produced a sequence remarkably similar to that in ancient bones from Tonga, American Samoa, and Easter Island as well as that from the living Araucana. The Easter Island sample differed from the ancient Chilean bird by only one base pair, making it nearly an identical match. All the pieces seemed to fit. "An ancient Polynesian haplotype persists in modern populations of Chilean chickens," the 2007 paper concluded. "In the 600 years or more since the introduction of European chicken these sequences have deviated little from their ancient Pacific ancestors."

A South American geneticist living in Australia, Jaime Gongora, was skeptical of the results. Gongora teaches at the University of Sydney but grew up with chickens in rural Colombia and recalls as a child wondering why Araucanas lay blue eggs. As Storey was analyzing her bone, Gongora was sequencing South American birds with hints of East Asian genetics, but concluded that the mixing could have taken place as late as the 1930s. When Matisoo-Smith and Storey shared their preliminary results with him, he was doubtful of the pre-1492 dates.

Gongora, a thickset man who might be mistaken for a Polynesian on the streets of Sydney, meets me at a Thai restaurant near the university. During our meal, he points to a shrimp dish we share. When we eat and drink, we take in the radioisotope carbon 14—also called radiocarbon—created when cosmic rays hit the earth's atmosphere. It pervades the sky, land, and water. Over time it loses electrons and its radioactivity. How far along organic material is in that decay process is what allows scientists to date a piece of charcoal, a seed, or a piece of bone with such accuracy. Oceans, however, mix deep waters that are very old with surface water that might be very young.

Radiocarbon date yourself after eating shrimp and you might appear to have added a few years as well as ounces after the meal. Scientists call this the marine reservoir effect. Gongora assumed that if the El Arenal chickens lived close to the sea, then the birds may have feasted on seafood and therefore their bones only appeared to date to pre-Columbian times. "If they were living with fishermen, then it is possible the chickens were eating the catch," he says. The paper by Matisoo-Smith and Storey did not take this effect into account.

Gongora and others, including Alan Cooper at the University of Adelaide in southern Australia, also were not persuaded by the genetic analysis. In blood samples from forty-one modern Chilean chickens, they found that the birds shared a haplotype common to European chickens. The El Arenal bone has a sequence that is the most common type found around the world. Some suspected the bone may have been contaminated with modern genetic material during the course of Storey's analysis, an ever-present threat in DNA sequencing. Hunt, the Hawaiian archaeologist, was on Matisoo-Smith's original team but was persuaded by Gongora's evidence.

The battle of the bones was on. Matisoo-Smith, Storey, and Chilean colleagues used isotopic analysis to determine that the El Arenal chickens had eaten primarily land-based foods like corn and not seafood. They deny that there was contamination and reject the analysis by Gongora and his colleagues. "No one has produced any data that contradict our result," Matisoo-Smith told me recently. But her team did back off claims linking the El Arenal birds to the modern Araucana chickens, which do appear to be largely a twentieth-century creation, as Gongora claimed. An independent team of molecular biologists has since asserted that DNA studies of modern chickens hint at two stages in their spread in the New World, supporting evidence of an early Polynesian Pacific and later Spanish Atlantic arrival of the bird.

Whether the chicken first arrived via the Pacific or Atlantic, the controversy put a new spotlight on the possibility that Polynesians reached the New World before Columbus. Recent genetic work shows that the sweet potato passed from the peoples of the Andes highlands and into the hands of Polynesians, who spread it west all the way to New Zealand. Wind and current studies suggest that a Polynesian canoe leaving Easter Island could reach the Chilean coast, ride the current north to Ecuador, and then cycle back west on the trade winds.

Historical records mention South Americans along the Pacific coast using sails and possibly manning floating trading stations set up just offshore. Words, tools, ritual objects, and vessels like the sewn-plank canoe used by tribes in the southern half of western South America have many similarities with those in Polynesia. And fowl have emerged as an important tool in reconstructing the audacious but poorly understood explorations of Polynesian seafarers as they settled one-third of the globe.

Chickens have thrived on the Hawaiian archipelago as long as humans. Island myths, many centered on Kauai, are filled with stories of devious roosters and beautiful women who shape-shift between hen and human. In one tale, a virtuous hen-woman named Lepe-a-moa takes on the evil and powerful cock of the king of Maui. The ruler's gullible opponents around the cockpit would gamble away their canoes, mats, headdresses, and even their own bones. But Lepe-a-moa slipped into the ring as the king of Oahu's contender and defeated the Maui rooster. "She tore him to pieces, until the battle was in a thick cloud of flying feathers," one version of the story relates.

The bird has long had royal and magical connotations in traditional Hawaiian culture. "Hawaii is a cockpit, on the ground the well-fed cocks fight," intones an eighteenth-century chant commemorating a victory of King Kamehameha, the cocks being the island chiefs. Archaeologists working here have uncovered remains of the bird in the houses of the ancient elite and rarely among the common people. Since the intrepid adventurers who settled the Pacific Ocean left behind no texts and few artifacts, these precious bits of cartilage provide welcome insight into an astounding era of discovery that scientists are only now starting to piece together.

One of the richest sources of ancient Polynesian chicken bones is just above a Kauai beach. Makauwahi Cave is a limestone sinkhole wedged in a crease between high cliffs that meet the Pacific Ocean and a long stretch of sand favored by sailboarders. Looming in the distance is Mount Ha'upu, its pointy ridge home of the god Ku and the goddess Hina, who gave birth to an egg that revealed a chicken, according to one ancient creation myth.

The cave is a giant pickling jar recording thousands of years of geological changes, biological invasions, and waves of humans that successively altered the landscape and encapsulate much of the island's history. The archaeologist David Burney discovered what may be the richest fossil site in the entire Pacific region by accident. In 1992, he followed the tracks of tourists to a rock wall punctured by a narrow crevice and found a vast oval of rock rising sixty feet high. Ever since, he has probed the thick mud underneath, pausing to give the actor Johnny Depp a tour before Captain Jack Sparrow leaped off the cave lip in a recent *Pirates of the Caribbean* film.

When I arrive at the site one breezy morning, Burney is at the bottom of a deep pit that is covered with a bright-green scum. Filling buckets with black ooze, he then hauls each by hand up the twenty-foot aluminum extension ladder and hands it over to a volunteer. The heavy muck is taken to the shady side of the natural coliseum and poured slowly into mesh boxes that sift for bones and shell. Burney, trim as his gray beard, invites me down the ladder. As I begin my descent, he calls up that I'm passing areas where they found a piece of glass or an iron nail bartered by the crew of a clipper ship, and then an area that was filled with large boulders deposited by a tsunami four centuries ago. These were nightmarish to remove but sealed the prehistoric layers from later disturbance. About two-thirds of the way down the ladder, he yells for me to stop. "This is where the chicken bones are," he says from down below, pointing at a dark layer of earth in front of my nose. "And we can be pretty sure they are not mixed with modern stuff. There is not KFC chicken in here at all."

Finding undisturbed and uncontaminated chicken bones that are centuries old is a remarkable feat. They typically don't last a day or two in most traditional villages. Dogs, rats, and other animals quickly devour what little is left. Insects and soil can eat up what remains of that. Burney explains that if the ground is too acidic, bone degrades although seeds fossilize. If it is too alkaline, bones will turn to fossils but vegetation degrades. Here, the alkaline limestone and the acidic groundwater cancel each other out. "This is the Goldilocks zone, perfect pH, just right," he says. "Which means both types are preserved. Basically, everything in here is preserved. It's like pages in a diary."

We clamber back up into the hot sun, which has just risen above the lip of the cliff, cross into the shade, and watch a retiree in a white tennis hat sorting through the mud in a rectangular box with quarter-inch mesh. "Don't do every last little snail, but every bird bone and every seed we want to keep," he tells her. "The biggest problem is that people try to screen too much at once," he says, turning to me as she uses water from a garden hose to dissolve the mud. When she's done, we hose off the thick black mud that coats our legs to the knees.

Burney ships the chicken bones he finds to the southern Australian city of Adelaide. The five-thousand-mile trip, which required weeks on the fastest clipper ship, now takes ten hours. The bones arrive at a building set in the middle of Adelaide's prim botanical garden. Until recently, extracting delicate strands of DNA from a long-dead sample was science fiction. Now researchers are pulling DNA from thirty-two-thousand-year-old algae dredged from the sea floor, eighty-thousand-year-old archaic humans, and horses that lived seven hundred thousand years ago. This technology makes human companions like the chicken important markers in understanding human movement around the planet.

To see how this is done, I pay a visit to the Australian Centre for Ancient DNA. Peggy Macqueen, a youthful, round-faced woman with cropped white hair and black clothes, greets me at the entrance door. She had already told me to shower and change into clean clothes before arriving. In the foyer is a small locker room for my camera, phone, notebook, and pen. The strict protocols are designed to protect the samples from the innumerable genetic codes—not just my own—clinging to my hands, my hair, and my very breath. Microscopic leftovers of an egg sandwich could confuse analyses of Burney's bones.

Macqueen takes me down the hall and up a flight of stairs to a changing room where we don white all-body suits, face masks, and gloves. We sit on a knee-high bench that divides the room and swing our legs over. Then she directs me to put on a second layer of latex gloves in the final staging area. Then she has me pull on a third set of gloves.

Finally we are ready to pass through an air lock and into the lab itself. As we enter, I feel a slight whoosh of air against the tiny bit of skin on my face that is not covered. The air pressure inside the lab is higher than the pressure outside to limit contamination. The room we step into is spare, with white walls, a counter, and two computer terminals. There is none of the usual lab clutter of half-filled coffee cups, clipboards, and gaping backpacks.

Macqueen guides me to a metal door in the corridor just beyond, which opens into a room the size and shape and temperature of a restaurant cooler. An entire wall of cubbyholes is filled with samples in little plastic Baggies and boxes that seem to be an inventory patiently waiting for a *Jurassic Park*–style revival of Siberian bison, Patagonian saber-toothed cats, and one of the world's largest mammalian carnivores, the giant short-faced bear, which weighed as much as thirty-five hundred pounds when it prowled American woods. All went extinct, likely at the hands of humans, about twelve thousand years ago. There is a whole section of birds and, on the other side of the cooler, a corner devoted entirely to dozens of chicken-bone samples from across the Pacific—Vanuatu, Easter Island, and four from Maukauwahi Cave that made the flight from Hawaii before mine. Chickens here qualify as something apart from birds.

Down the hall we enter a chamber of blowers, like at the tail end of a car wash. The spacious room beyond is where Macqueen cleans cartilage to remove any chemicals on the outer surface. Then she cuts a section and puts the sample in a Mikro-Dismembrator, a boxy copy-machine-sized device that, within ten seconds, pulverizes a bit of bone into less than an aspirin's worth of powder. This she adds to a solution of enzymes designed to break down any organic material and turn the long strands of DNA into smaller and more manageable lengths. After baking it overnight at 130 degrees Fahrenheit in an incubator, Macqueen centrifuges the remaining goop at ten thousand rpm for five minutes to separate out the various components. Then she adds molecular-grade water, centrifuges the sample once again, and incubates the result for an hour or so. "You end up with nice clean DNA in nice little tiny degraded fragments," says Macqueen, her voice slightly muffled by the mask. "And then you amplify it."

We move into the third and most restricted part of the lab, a series of glass-walled rooms with windows looking out over the botanical garden's roses. The yellow-painted room, with space for only a desk and chair and a long and narrow counter, is designated solely for an-imal DNA material. This is where Macqueen uses a technique called polymerase chain reaction—PCR—to make thousands and even millions of copies of those nice little tiny degraded fragments. Even tiny surviving portions of DNA in a badly degraded bone can then be analyzed with relative ease. This obscure technology, developed in the early 1980s, won its inventors the Nobel Prize in Chemistry for opening the door to reading the genetic code embedded in all living things. "The last thing you do is add the DNA to a test tube, clean it, and put it in the machine," she says. "You close the door, and from then on it is amplified DNA."

The next job is to clean, clean some more, and clean again. The desk, the counter, the chair, everything must be wiped down with disinfectant to ensure that as little stray DNA remains in this room as possible. The invisible strands from one sample could contaminate the next sample that comes through the door, so such precautions are critical. She spends more time cleaning than conducting the actual work, and she must work efficiently and in isolation. You can't leave the lab for a bathroom break without going through the entire prepa-ration protocol all over again, and tight restrictions limit how many times a day you can reenter. All food and drink, including the coffee and chocolate she likes, are forbidden. Even the building is set apart from the rest of the campus, in the middle of the arboretum, to create yet another barrier from contamination.

Once the DNA is safely copied, she might store some in a freezer and carry the rest in a tube to the biology building, a ten-minute walk away. Until recently, she had to hurry on hot days, since high temperatures could degrade the sample. But now she uses a new chemical that stabilizes the precious material and she can stroll through the gardens on her way to campus. Still, the work is demanding and often frustrating. "Getting mitochondrial DNA out of these old bones is a struggle," she says, reflexively wiping down a counter for a second time. "You have to decide what samples are worth a go."

We dispose of our protective wear and Macqueen grabs her coat and leather backpack to take me to the lab's offices across campus. At a café in the garden, she tells me that she grew up on a farm in northern Australia with cattle and chickens. Later she worked in Southeast Asia on development programs designed to improve poultry for the rural poor, where they introduced the large, heavy-breasted chickens favored in the West, the birds that resulted from combining European and Asian breeds in the past century. Locals in Laos and Cambodia preferred their smaller and scrawnier birds that lay fewer eggs. "They thought their village chickens tasted a lot better, and they were right," she says. The project failed and she moved on.

What struck her during her time in Southeast Asia was that the bird was something more than just food; it was a multipurpose animal for amusement, religious ritual, and gambling. In ancient Polynesia, the bird's meat and eggs were of course a source of fresh food on long voyages, although, like many tribal peoples of Southeast Asia, Polynesians never appear to have been major egg eaters. Instead, they used the bones for sewing and tattooing, the feathers for decoration, and the roosters as an excuse to gamble. Cockfighting was associated with religious ritual as well as entertainment. Ruaifaatoa was the god of cockfighting in Tahiti, where tradition held that chickens were made at the same time as humans.

Macqueen drops me off at the office of Alan Cooper, who directs the center and helped define the strict protocols for ancient DNA research when it began in earnest a dozen years ago. The New Zealander has feathered hair and a boyish face and speaks quickly, as if he's perpetually late for his next meeting. "The Pacific is a basket case," he says. For archaeologists, the vast region is no paradise. "There are not enough human bones, preservation isn't good, and it is a hard place to work." Animal remains offer a better alternative way to chart the movement of humans, since they are more common than those of humans and don't upset locals fearful that the tombs of their ancestors will be desecrated. Not all Polynesian colonists carried the same plants and animals. No prehistoric pig or dog bones have been found on Easter Island, for example. "And rats," says Cooper, "are a pain in the ass." They could have stowed away on boats that traveled back and forth among islands, and their relentless comings and goings make their genetic signature even more scrambled than that of the chicken.

The bird, he adds, is proving the best way to reconstruct human colonization of the Pacific. Any chicken remains found west of Bali, the eastern edge of the red jungle fowl's home range, are signs that humans carried the bird in boats. By pinpointing combinations of DNA sequences, haplotypes assigned letters like A, B, C, and D, molecular biologists can connect the dots of this movement from west to east across the Pacific. Old bones like Burney's are rare, so the search also embraces DNA samples from modern chickens that may still contain pre-European genes. Cooper's team extracted DNA from 122 modern and 22 ancient chicken samples from across the Pacific, and discovered that the vast majority, both ancient and modern, shared the same haplotype—designated as "D."

The ones that did not share the D group—mostly of the E haplotype—came from more populated areas that have a longer history of contact with the outside world in recent times. Samples that Macqueen retrieved from more remote villages of Vanuatu, six hundred miles southeast of the Solomon Islands, for example, were of the D type, as were four ancient bones found by Burney in Maukauwahi and six samples from a prehistoric site on Easter Island. This particular haplotype combines four specific genetic sequences that allow researchers to simulate various migration routes of chickens—and therefore people—from west to east. "Polynesian chickens may be one of the few examples where ancestral genetic patterns can still be observed in a domesticated species," Cooper and colleagues state in a 2014 paper. These birds also may still harbor some of the world's last undisturbed genetic material from precolonial chickens.

Cooper's team found that people took two primary routes. The first stretched from New Guinea to Micronesia on the northern end of Polynesian expansion. Once they reached Micronesia, a scattering of small islands, they stayed put. Some modern chickens on the isolated Micronesian island of Guam—fifteen hundred miles due east of the Philippines—contain a unique form of haplotype D not shared with other Pacific islands, while other Guam birds are linked to a subgroup found in the Philippines, Japan, and Indonesia. Micronesia was out of the mainstream of Polynesian expansion. By contrast, the ancient chickens of Easter Island and Hawaii appear to have moved across a southern route that passed through New Guinea, the Solomon Islands, and then east. The ancestors of the birds found by Burney came via Melanesia, a vast island group that includes the Solomon Islands and stretches east to Fiji.

Such finds combined with new dating of bones, pottery, and artifacts from archaeological digs across the Pacific suggest that humans took big leaps and then paused. The first and longest halt was at the Solomons, east of New Guinea, separated from the next set of islands by more than two hundred miles. By 1200 BC, when Ramses the Great was on Egypt's throne and the first chickens clucked in Egypt, early Polynesians were working their way into the open ocean as far as Fiji. There they remained until Samoa and Tonga were settled, about 900 BC. There was another long hiatus in successful voyages east, as explorers faced the daunting void in the middle of the world's largest ocean, until two millennia later, around the eleventh century AD, when settlers colonized the Society Islands in the central South Pacific. The last burst of movement took place as late as the thirteenth century—several centuries later than once thought—when Polynesians finally made landfall on the Hawaiian archipelago and Easter Island.

The origin of the culture that birthed the Polynesian expansion remains murky. Archaeologists call it Lapita, a name that comes from a site on the island of New Caledonia, halfway between New Guinea and New Zealand, that was dug in the 1950s. Since then archaeologists have found hundreds of sites with similar remains scattered throughout the region. These include stone axes, sweet potatoes, sugarcane, gourds, taro, bananas, bamboo, turmeric, pigs, rats, dogs, and chickens. There were also stowaways like snails. The people lived in houses on stilts, baked food in earthen ovens, and fished. But how these objects, plants, animals, and traditions came together, where these people came from, and how and when they spread to every major island remain hotly debated.

One view is that the Lapita people were rice-farming immigrants in China who crossed the sea to Taiwan, moved south to the Philippines and Indonesia, bypassed the inland farm peoples of larger islands like New Guinea, and then hit the open ocean. Others argue that they emerged from the mass of islands between Indonesia and the Philippines and then spread east into Melanesia and into the Pacific. A third group argues that Polynesians didn't travel to Melanesia at all but were indigenous to the area. The disagreements, which often hinge on arcane linguistic theories, reveal just how little archaeological material exists.

Cooper's team has intriguing hints that the domesticated chicken first joined the Lapita package at an early stage, in the Philippines. All four of those unique genetic subgroups in haplotype D were found in one modern chicken from the little southern Philippine island of Camiguin. A Filipino graduate student has five hundred modern samples and ten ancient ones from around the archipelago, which he hopes to connect to birds on the mainland of Southeast Asia or in Indonesia. That could help pinpoint an origin of the mysterious early Polynesians.

Since haplotype D is associated with game-fowl breeds from Japan to the Philippines to India, cockfighting may have been a major factor in the spread of the chicken beyond its home turf. The possibility that the chicken's ability to fight may have trumped its meat-and-egg capacities in ancient times intrigues scientists. No people on earth have bred the gamecock as avidly as Filipinos. Since the sixteenth century, the island chain has been known as the home of some of the world's greatest varieties of fighting birds. Even today in the Philippines, cockfighting remains as central to traditional life as bullfighting is in Spain. Forgotten or disdained in much of the industrial world today, the sport was likely a catalyst in the chicken's spread around the world.

5.

Thrilla in Manila

So lively an Emblem of true Valour is the well bred-Cock,
that he is not to be parall'd amongst the many Creatures
which the wise Creator of all things has been pleased to
make Man the Lord and Master of.

—Robert Howlett, *The Royal Pastime of Cock-fighting*

The World Slasher Cup is the Super Bowl of cockfighting, a five-day series of 648 matches held in a coliseum in downtown Quezon City in metro Manila. Outside the sleek chrome entrance, a thirty-foot-high inflated rooster sways in the hot breeze, advertising a formula feed. Next to the event poster featuring two cocks in combat is a bill announcing the fiftieth Miss Philippines contest, displaying a beautiful woman in a teal-blue low-cut dress. Ice Capades just finished a run and Dionne Warwick will be coming soon. But for now, the twenty-thousand-seat arena is devoted to the Filipinos' favorite traditional pastime and humanity's oldest spectator sport after boxing.

In 1975, this was the site of the famous bout pitting Joe Frazier against Muhammad Ali for the title of world heavyweight champion, the legendary Thrilla in Manila. In the annual Slasher Cup competition, two gamecocks, each armed with a long, curved steel spur, will battle to the death. Big screens make it easy to watch the combat even from the upper tiers. When I arrive there are four men in the ring, two of them calmly squatting, each with a cigarette between his lips and a chicken between his legs. The other two are referees. Thousands of spectators, all men, are standing and shouting, making distinctive hand gestures to one another around the vast space. The noise is deafening.

Suddenly, the squatting smokers release the birds, and the roosters approach each other at a wary angle, reptilian hackles rising like rainbow-colored umbrellas from their necks. As they explode forward with the speed and aim of heat-seeking missiles, the clamor outside the ring abruptly halts. Feathers, legs, and a flash of steel fill the screens. The only sound is the vibration of pounded air from hard-flapping wings. In less than a minute it is over. The white-feathered victor sends up a triumphant crow next to the still body of its dead opponent. Losers pay up their bets in a rain of folded peso notes as the loudspeaker blares the pop tune "Eye of the Tiger."

"Here in the cheap seats, they are betting ten to a hundred dollars a match," says Rolando Luzong, my cockfighting guide. We are sitting halfway up, where the crowd thins out. "But there in the *preferencia*"—he points at the VIP bleachers next to the ring—"they are betting a thousand to ten thousand dollars." Hundreds of thousands of dollars are passing hands with each of the 648 matches. The owner who scores the most points at the end of the meet will get a check for thirty-five thousand dollars and a laudatory article and picture in the sports sections of the Manila papers. The real money, Luzong tells me, goes to breeders profiting from selling individual winning birds to sire lines of future champions, and to the companies hawking steel blades, supplements, shampoos, and special feed for the tens of thousands of gamecocks throughout the country and beyond.

Luzong makes his living from cockfighting, not as a gambler or breeder but as a journalist, website developer, public-relations specialist, and industry consultant. In his bright-red shirt advertising Thunderbird feed, he's an unabashed promoter of a sport that is illegal and reviled in half the world. Middle-aged with jet-black hair, a deep slouch, and a belly as round as his face, Luzong came late to the business. Pushing thirty, he landed a job writing for a game-fowl magazine and discovered that he was less interested in the gambling aspect than in the peculiar names of bird varieties—Roundhead, Butcher, Sweater. Luzong later spent a decade at Manila's Roligon Mega Cockpit as public-relations chief and then as general manager. The pit is the world's largest, an enormous stadium near the airport.

As we sit through another match, he explains the rules. First, potential rivals are compared and weighed to ensure an equal match. Then the artificial spur is fitted and tied like a boxer's glove to each bird's left foot—the natural bony spur is amputated while they are young—and carefully wrapped in a protective covering. The ring manager and the game-fowl owners agree on a betting figure for the winner, and each fighter is assigned one side of the ring. These are the inside bets. The bird most favored to win is on the side called Meron (with a hat) and the other is on the Wala (without a hat). This comes from the tradition of one owner wearing a hat while the other goes bareheaded to ensure that spectators know which owner's bird they are betting on. In the twenty-first-century Smart Araneta Coliseum, the words are spelled out on large electronic signs hanging above the boxing-match-sized ring.

Only four people are in the ring at one time—two handlers, a referee, and an assistant referee. In the two thousand or so village pits around the Philippines, a handler usually is also the owner. In the high-roller world of cockfighting, however, specialists are hired. Once in the ring, they typically work the cocks into an angry and agitated state using another bird brought for this express purpose. Spectators then have a chance to judge where to put their money. Once the inside betting is completed, betting managers around the ring spread their arms—they are called Kristos in this overwhelmingly Catholic country because of their crucifixion-like stance—and solicit bets from the crowd.

This was the din I was hearing just before the first match. That noise builds as individuals bet with the Kristos or among themselves using well-known hand signals that specify the number of pesos being wagered. Meanwhile, in the ring, the blades are unwrapped and the birds are released. All betting ceases the moment they engage. "The fastest one I've seen lasted eight seconds," Luzong says as a rooster dives on its opponent in the ring far below. If a match lasts ten minutes, it is considered a draw. But most are over in a couple of minutes. If a cock goes down for the count, the referee will pick them both up. If one pecks but the other doesn't, then the pecker is the winner. If both die, the one with the most pecks wins.

In nearly every case, only one bird leaves the ring alive. There is not much blood to be seen, even with the big-screen videos providing up-close action shots. Occasionally, a cleaning crew mops up dashes of crimson. Dead losers are unceremoniously carried away while injured victors are taken to a makeshift triage station in a room behind the bleachers where veterinarians apply painkillers and stitch up those birds that might one day enter the ring again. But it is a rare bird that returns to fight more than twice. Serial winners are highly sought after to sire a new line, and the lucky owner can expect fame and fortune. "If you win three times, you will have as many girls as you want," Luzong tells me.

Until a game fowl is a serial winner, however, it has no name. If your bird wins, you get a single point; a draw gets you a half point, and a loss none. The contest is about the human rather than the animal. Like the metal contestants in shows such as *BattleBots* or *Robot Wars* that were popular a decade ago, the bird is simply an extension of its owner.

Among the several thousand spectators that evening, I see only one woman. She is near the front in designer glasses, a tight T-shirt, and fashionable jeans and clearly is a confident and experienced gambler. A dozen Hawaiians in matching yellow shirts emblazoned with the words "Summit Farms" sit together in the bleachers. In the *preferencia* are a scattering of pale Americans in Dockers pants and polo shirts along with Filipino men in button-downs and expensive shoes. Luzong watches my gaze. "They are senators, congressmen, presidents of corporations, and big businessmen there," he says. "A lot more than gambling is happening right now. Business deals and

political decisions are being made. Powerful men are building cama-
raderie. Being here can really help your political career."

The cup is the high-end of global cockfighting. To enter a single
bird in the competition costs $1,750, more than half a year's salary
for the average Filipino. Wealthy owners often have more than just
the money; they have dedicated farms and full-time trainers caring
for hundreds of fowl that could sell for well over a thousand dollars
apiece. On top of that, they purchase expensive feed and supplements. "Our sales last year were eighty million dollars," says Luzong
of Thunderbird. "And we're just starting to promote medicines."
Vaccines, antibiotics, vitamins, and supplements are all part of the
modern game fowl's life. Traditional methods of revving up your bird
for a fight, like slipping cayenne up its anus, have given way to pricey
steroids and other powerful drugs.

Like American baseball or the Tour de France, modern Filipino cockfighting is caught in a tangle of corporate sponsors and
performance-enhancing drugs. The brightly lit concession stands,
the blaring canned music, and the rows of clean toilets in the rest-
rooms give the event a depressingly modern feel. Still, the people in
the cheap seats are the working-class men whom you would find at
any Canadian hockey game, British rugby match, or Brazilian soccer
contest. The real draw, though, seems to be in the gambling outside
the ring rather than in the combat inside.

Luzong insists that cockfighting is a much less corrosive form of
gambling than what takes place in the flashy casinos springing up
across Manila. "Here, you only choose between two birds; you have a
fifty-fifty chance," he says as another shower of peso notes sprinkles
through the air at the conclusion of a match. "You can cancel your bet
before the roosters are released. When the birds are released, there is
no human intervention. And you can leave whenever you want."

There are, of course, other stories. There is the young villager who
spent all his hard-earned money saved in an overseas job while in a
cockfighting stupor. A family lost its small plot of land because of an
ill-placed bet. And children may be treated only half as well as a father's prized cocks that enjoy a carefully planned diet, constant atten-

tion, and even piped-in music and air-conditioning. For the millions packed into Manila's sprawling slums, cockfighting offers a fast way up the socioeconomic ladder or a quick tumble into indigence.

Such high stakes have an extraordinarily long history. Filipinos raise "large cocks, which from a species of superstition, they never eat, but keep for fighting purposes," Antonio Pigafetta wrote five centuries ago. "Heavy bets are made on the upshot of the contest, which are paid to the owner of the winning bird."

Pigafetta was among Ferdinand Magellan's crew when they became the first Westerners to cross the Pacific Ocean and appear on Filipino shores. Cockfighting was popular in sixteenth-century Spain, but it was already an obsession in the Philippines. When the hungry and exhausted crew arrived in 1521, locals paddled out in boats to offer "sweet oranges, a jar of palm wine, and a cock, to give us to understand that there were fowls in their country," Pigafetta recalled. That bird likely was raised to fight rather than to be eaten.

Magellan quickly wore out Filipino hospitality by embroiling the expedition in a local conflict. He was speared to death. Pigafetta was one of only eighteen men on board the sole remaining ship, the *Victoria*, as it limped back to Spain. That might have been the inglorious end of the Spanish in the Philippines, but as Pigafetta's vessel worked its way through the Indian Ocean and then up along the West African coast, Hernán Cortés was finishing off the great Aztec Empire in Mexico. A decade later, Francisco Pizarro and his soldiers took on the Incan Empire in the Andes of South America. Spain soon was in control of vast stores of gold and silver from the New World, and enormous mines worked by Native Americans produced astonishing wealth at a horrifying human cost.

Spain needed a stronghold close to the markets of China and the Southeast Asian islands like Sumatra, which had the spices, silks, and other luxuries craved by increasingly prosperous Western consumers. China was closed to foreigners. Their merchants, however, wanted New World silver and gold. The Philippines, sitting astride the major sea routes in the western Pacific, proved a strategic military and economic base for the wealthiest European nation to consolidate its position in the East. The first three attempts at colonization ended disastrously, including one organized by Cortés himself. But the fourth succeeded, and by the end of the sixteenth century, Spain had a commercial treaty with China and a well-fortified port in Manila. Spanish galleons laden with New World silver mined by Native Americans sailed from Acapulco to Manila. There they could trade with Chinese merchants. That lucrative trade lasted for nearly three centuries and depended heavily on Spain's control of the Philippines.

Governing a colony consisting of hundreds of islands with dozens of ethnicities half a world from Madrid posed a challenge to the distant Spanish rulers. Copying what they had done in the New World, friars and civil administrators in the 1600s set up a feudal-style system across the colony. Scattered populations were concentrated in towns so that natives could be monitored, taxed, and used as a labor force. The ingrained passion for cockfighting in the Philippines offered Spanish administrators both an important revenue source and a way to control the population. Spiritual wealth through Christianity and earthly wealth through cockfighting proved key levers in drawing the native peoples of these islands into towns and generating the tax revenue that made this Asian beachhead pay for itself. Even today, almost every Filipino village has three main structures—a church, a town hall, and a cockpit.

Visiting foreigners were appalled by the enthusiasm the locals had for the sport. "The sight is one extremely repulsive to Europeans," sniffed one nineteenth-century German traveler. "The ring around the cockpit is crowded with natives, perspiring at every pore, while their countenances bear the imprint of the ugliest passions." He was shocked by the "incredibly large sums" wagered, and blamed the country's rampant theft, highway robbery, piracy, and any number of other ills on "ruined gamesters."

The passion for cockfighting spread via the Acapulco trade to Spain's empire in Latin America. Though the sport had been popular in Spain since Roman times, it was the Filipino fever that swept to what is now Mexico, Colombia, and Venezuela. Filipino fighting birds were exported all over the world. According to some historians, the wandering fowl of Florida's Key West arrived from Cuba but came from Spain via Manila and the gamecock trade.

As early as the 1700s, the Spanish government regularly sold licenses for Filipino cockpit operations to the highest bidder. Cockpit licenses combined with game-fowl sales may have generated more money in the nineteenth-century Philippines than tobacco, the colony's single most important export. In 1861, Madrid's revenues from these licenses topped a hundred thousand dollars annually. That year, a special ordinance from Madrid permitted cockfights on Sundays and feast days, after the conclusion of high mass and before sunset. "It being a vice, the laws permit it only on Sundays and holidays!" one outraged American evangelical complained later. "This condition is a heritage of Spanish times when the friars and priests owned many of the cockpits as well as one tenth of the improved land," he charged. "They wanted their subjects to work during weekdays and to come into the *poblacion*"—the center of a Filipino town—"to early mass on Sunday morning, and spend the remainder of the day at the cockpit, gambling away all the money they had; getting themselves deeper into debt to the Spanish overlords."

That policy ultimately backfired. José Rizal, the guiding spirit of the Filipino rebellion against Spain, was no fan of the sport. In his moving 1887 book *Touch Me Not*, he compared it to opium smoking. Rizal was a European-educated poet, sculptor, doctor, and polyglot who despised what he considered a backward tradition. "To the cockpit went the poor man to risk what little he had in order to make money without working," Rizal wrote, "as well as the rich man who went to amuse himself with the money left over from his parties and the thanksgiving Masses."

He recognized, however, that these regular events that took place in virtually every town across the colony's hundreds of islands could serve as both inspiration and a cover for revolutionary activities. Cockfights provided Filipino men a place to meet freely and regularly in large crowds, while the combatants offered an example of the courage required to take on the behemoth that was Spain. Rizal recalls a moment when a favored rooster loses a match to a scrawny underdog. A roar goes up from the crowd. "So it is among nations," he wrote. "The small nation that achieves a victory over a larger one tells and sings of it forever after."

Spanish officials executed Rizal in 1896, and he became a martyr to the cause of independence. Cockfighting is forbidden by law on the anniversary of his death out of respect for that sacrifice. Three years later, the country that sought freedom from one empire found itself embroiled in a bloody guerrilla war with another behemoth. The United States annexed the Philippines before the nascent nation had a chance to assert its independence. Cockfighting was seen by the new conquerors as one more reason to deny Filipinos self-determination. "How can a people who allow themselves to be known by the barbarous sport of cockfighting be allowed to govern themselves?" asked W. A. Kincaid, the president of the Philippine Moral Progress League.

Kincaid was an American lawyer who held meetings around the provinces "with the object of inculcating a hatred for the vice," according to a 1900 report to the U.S. Secretary of War. This was a time when the sport was still widespread, popular, and mostly legal in the United States itself. Filipino politicians who wanted to land on the good side of their new rulers responded to the shifting winds. "Cockfighting has spread to such an extent that it is no longer mere sport or amusement for the people, but now constitutes a veritable vice followed by . . . crimes, the ruin of families, and an inexhaustible opportunity for exploitation," wrote Isabelo Artacho, the U.S.-installed governor of Pangasinan province, in the same report. "A law should be enacted restricting this form of gambling to the greatest possible extent at once."

Cockpits quickly became magnets for American Protestant missionaries looking for a crowd and for converts. Meanwhile, the U.S. government distributed English-language textbooks criticizing the tradition as a crude, outdated, and dangerous activity. The ultimate weapon in the fight was the favorite American pastime. "Baseball is a

tremendous factor in educating the vigorous Philippine lads not only
in the play they instinctively yearn for, but in the teamwork which
is needed to eliminate some of the unfortunate features of the older
civilization," one charitable organization noted in 1916. The most unfortunate feature, of course, was cockfighting. Japanese invaders arriving in 1941 also tried to stamp out the practice, which they viewed
as barbaric. These efforts at moral improvement had little effect.

As other nations over time criminalized the sport, Filipinos de-
fiantly kept it as their own national pastime. When the American
author Wallace Stegner visited in 1951, baseball had not cured the
nation of its long-standing tradition. "You don't know Filipinos until
you have seen some little fellow who has trained a chicken for months
put it into the ring against another's rooster," he wrote. "He bets everything he owns, steals his wife's savings, sells his children's shirts
to raise a peso. If he wins, glorious; if in one pass his rooster gets its
throat cut, then you will see how a philosopher takes disaster." That
stoicism remains part of cockfighting's tenacious culture; no loser
that I witnessed at the World Slasher Cup—and there were thousands
every few minutes—expressed anger, frustration, or regret.

The tradition's staying power outlasted even the dictator Ferdinand Marcos, who imposed martial law and jailed dissidents as he
looted the country for two decades starting in 1965. Fearful of large
gatherings of men who might plot against his rule, Marcos attempted
to restrict the sport in the mid-1970s. Under pressure from his advis-
ers, he backed down. His wife, Imelda, is said to have once sponsored
a cockfight in which the winner drove away in a new Mercedes-Benz.
The only successful ban imposed on the sport is a law forbidding
cockfights and liquor sales on Election Day and the commemoration
of Rizal's martyrdom.

By the 1990s, the growth of cities like Manila and distractions like
satellite television and shopping malls threatened to push cockfighting,
with its rural roots, into obscurity. Then, in 1997, the central government passed cockpit-licensing control to local authorities. "That
opened the floodgates to more cockpits and more fights," says Luzong.
"There is both good tax money and under-the-table money." Although

elected officials can't own a cockpit license, their friends and relatives can. The flagging sport rebounded as local governments took advantage of the revenues. Two million Filipinos directly benefit from cockfighting through the rings, hotels, restaurants, and shipping companies, Luzong says. "If you stopped it, there would be big economic problems," he adds. "It probably employs more people than the government."

That is a wild exaggeration, no doubt, but looking at the thousands of eager gamblers in the Smart Araneta Coliseum, it does not seem ludicrous. And the business is one way to draw wealthy foreigners. Gambling-mad Asia is much richer today than at any time in the past. Chinese tourists crowd casinos in the humid Cambodian towns that border Thailand. The island of Macao touts itself as the "Monte Carlo of the Orient," and even Muslim Malaysia boasts a massive mountaintop gaming complex with more than six thousand hotel rooms. Cockfighting enthusiasts don't have the same political or economic muscle as developers with expensive resorts.

The modern casino industry sees as a threat any game that can be played without need of hotels, restaurants, and malls. Collecting revenues from remote village cockpits can be difficult for governments. The contemporary Western view of cockfighting as a barbaric practice also makes it a hard business to defend, and animal-rights organizations in South Asia are slowly gaining influence. As a result, cockfighting in the region where the chicken was first domesticated is in a defensive crouch.

The Philippines today is to cockfighting what Switzerland is to secret bank accounts, a safe haven where you can operate without government interference. Wealthy Malaysians and Indonesians come here to gamble, and Americans come to sell their fighting birds. Cockfighting in the United States is against the law in every state, but it exports more game fowl than even the Philippines. "Most Americans are breeders who come here to promote their birds," Luzong explains, nodding toward the VIP bleachers. "There are fewer diseases there. And the birds are tougher." Some sources say that American birds made their debut in the Philippines as early as the 1920s, brought by U.S. soldiers as American officials were attempting to stamp out the sport. Filipino birds are easy to spot, since they resemble red jungle fowl, while the American varieties are typically larger and a single color.

The United Gamefowl Breeders Association in the United States claims thousands of members, who raise hundreds of thousands of gamecocks. Since they can sell for a thousand or even twenty-five hundred dollars, it is a multimillion-dollar business. Animal-rights activists claim the organization collects money to defeat anticock-fighting legislation, a charge that association representatives deny. Luzong says that about one hundred Americans are in town for the World Slasher Cup, and most are breeders more than gamblers. He ducks my request for an introduction, and the few I approach give me a wide berth. They have good reason to be fearful. Police arrested game-fowl breeder Wally Clemons and seized two hundred roosters on his Indiana farm a few years ago after he gave an interview with *Pit Games,* a Filipino cocker magazine. American cockers understand-ably like to keep a low profile.

In the Philippines, by contrast, there is little organized opposition to cockfighting. The country's Animal Welfare Society steers clear of the issue. Any lawmaker's attempt to include chickens in cruelty laws is quickly squelched. In 2008, a small demonstration of people, in-cluding one man wearing a chicken suit, gathered outside the Araneta to protest animal cruelty at the World Slasher Cup. They were quickly hustled away, but Luzong was alarmed, and spearheaded a campaign using print, television, and Internet media to put cockfighting in an unassailable position.

Warming to his argument, he dismisses Western animal-rights activists as simply the latest version of morality-pushing imperialists from a century ago. Like bullfighting in Spain, he argues cockfight-ing is an essential piece of national identity. The sport has survived greedy Spanish administrators, hypocritical U.S. interlopers, ruthless Japanese troops, and the country's own thieving dictators. Unlike Western factory chickens that are pumped with feed and slaughtered within two months, game fowl are highly valued and well treated and live for at least two years before they enter a ring. "Then they travel in an air-conditioned Suburban and fight in an air-conditioned arena," he adds, slapping the seat in front of us. "They are probably the luckiest birds in the world."

Some 15 million game fowl die annually in Filipino rings. Winners typically carry the dead home for the family dinner. Those killed during the World Slasher Cup are buried in a mass grave. To Luzong, the competition is no different from dog shows. "Dogs are trained to do tricks. Roosters are trained to fight. They are both trained to win." After all, he says, roosters have an instinct to kill, and like humans, they are natural warriors who like to compete and win. When I suggest that the killing trait is one selected by humans—after all, red jungle fowl more often run than fight—he shifts gears. Cockfighting channels aggression in a less destructive manner than war, he says quietly. "You can see we are very loving people." He opens his arms to take in the scene. "Maybe it has made us milder because we are able to let the birds do the bad things for us."

Luzong looks at his watch. It is past 10 p.m. and I sense he's eager to get home to his family. They have no interest in cockfighting. "My children, they only see a chicken already cooked." He sighs, heaving his bulk out of the small plastic seat. "One time I slaughtered one and they couldn't bear it." We work our way through the crowd and into the deserted lobby. Methods, traditions, rituals, and specific beliefs differ, as do the continents, languages, and centuries. But it is all the same. The chicken does the bad things for us. It dies to save us from ourselves.

The rooster's comb is the bird's most distinctive feature. It serves as a heat radiator, a warning to potential rivals, and a come-on to hens. In the male red jungle fowl, the crest is red and serrated and called a single. Among chickens, however, there is an extraordinary variety, including the rose, the cushion, the buttercup, the pea, the walnut, the V-shaped, and the silkis. Each has its own specific history. The flamboyant buttercup mutation can be traced back to a breed developed by a medieval European royal family that ruled Sicily and

Normandy. The smaller pea comb is practical in cold climates, since it retains more heat. Combs can come in gray and bright blue.

Leif Andersson at Uppsala University realized that this physical feature, so altered by human selection, could offer clues to the bird's early domestication. Roosters with rose combs, for example, are known to have more difficulty fertilizing females than those with single combs. Andersson wondered why people would pay that steep price simply to alter the bird's headgear so that it looked like a little French beret. He and a team of nineteen collaborators used new genetic tools to identify the actual alleles involved in making combs. Alleles are those parts of a gene located in a specific position on a chromosome that is, in turn, a single piece of coiled DNA. Andersson and his team discovered that to create the rose comb, a gene jumps to another location, which disrupts the mechanism that ensures healthy swimming sperm. The study could conceivably help researchers understand low fertility in human males.

The rose comb's advantage is that it is more difficult for a rival to grab in the cockpit. Andersson realized that the wattles dangling from the cheeks of a rooster tend to be smaller in game fowl than in red jungle fowl or other domesticated varieties. The research provided direct genetic evidence that humans have long altered the chicken so that it is more likely to win in a fight—even at the cost of low rates of reproduction.

Red jungle fowl will fiercely defend their hens and chicks against larger adversaries, but the wild bird is just as likely to flee when confronted with danger. Breeding game fowl therefore required selecting for those birds that would stay and fight and had a physical advantage over their rivals. Some scholars speculate that cockfighting began in South Asia as a religious practice. A clan or village may have pitted its sacred roosters against another group's bird. In northern Thailand, for example, the *faun phi* ceremony honoring ancestral spirits entails cockfighting of a religious nature that may reflect ancient practices.

If cockfighting was a major driver in domestication, then the chickens' spread throughout South Asia and then to other parts of the world may be firmly tied to the sport. Today's American game-fowl breeders likely had ancient counterparts who traveled long distances with their valuable animals, bringing not just the chicken but the gambling practice to other societies.

One of the earliest recorded cockfights took place in China in 517 BC. The match was held in Confucius's home province of Lu during the philosopher's lifetime. By then, cockfighting was already a royal sport with elaborate protocols; the chicken had been in central and northern China for at least nine centuries. The birds were outfitted with metal spurs, and one was sprinkled in mustard to make it slicker and more difficult to grab. The match between two rival clans ultimately led to a war. "Thus from which disaster was born started in the foot of a cock," an ancient text relates.

The earliest unequivocal evidence of cockfighting in the West comes from this same era. In a tomb just outside Jerusalem, excavators found a small seal that shows a rooster in a fighting stance that was owned by Jaazaniah, who is called "the servant of the king." A man of the same name—Jaazaniah, the son of the Maakathite—is mentioned in both the biblical book of Kings and Jeremiah as an army officer after the Babylonian attack on Jerusalem in 586 BC. That battle ended with the leveling of Solomon's temple and captivity for its elite in Babylon. The artifact seems to date from this period. Another seal with a fighting cock belonging to Jehoahaz, named as "the son of the king," may come from the same era, but its provenance is unknown.

Philistines in the nearby port of Ashkelon on the Mediterranean coast at that time were likely breeding fighting cocks, since excavators have found numerous rooster remains with well-developed spurs. By then, they were making use of the hens as well, since their bones reveal that they were using the high levels of calcium necessary to produce large numbers of eggs.

Cockfighting retained a quasi-religious role in both China and the West in the early centuries before Christ. A Chinese Daoist work from about the fourth century BC tells the story of the time and effort involved in raising fighting cocks for the king, in a kind of parable of patience in developing spiritual readiness. At the same time, in ancient Greece, Athenians gathered in the theater dedicated to Dionysus to witness cockfights as a reminder that two battling roosters inspired Greek troops before they fought and beat a larger Persian force. The birds adorned the high priest's chair at the sacred theater dedicated to the god of wine and song.

Over time, spiritual combat transformed into a more secular event, perhaps as bullfighting began as a ritual and gradually turned into more worldly entertainment. In the early centuries AD, cockfighting was a regular pastime for many villagers, soldiers, and aristocrats in the Old World's two great empires, Rome in the West and the Han dynasty in the East. The sport became a place for men of many classes to meet, take financial risks, and watch male animals demonstrate raw courage.

By the early nineteenth century, pitting two roosters in a fight to the death was almost universally practiced, except in western and southern Africa, where it oddly never took root. In India, British officers matched their gamecocks against Muslim princes'. An English visitor to China noted in 1806 that a thousand years after the Tang dynasty, "one of their most favourite sports is cockfighting," adding that it was as commonly practiced among the upper classes as it was in Europe. There were many regional variations; some cockers used long spurs, some short ones, and others fought the birds with their natural spurs. But cockfights were typically places where men—and they were almost always just men—could rub shoulders regardless of race or class. A European visitor to early-nineteenth-century Virginia marveled that black slaves bet as avidly as their white masters at a match.

No one enjoyed the sport more than the English, who likely were pitting their birds before the arrival of the Romans and possibly absorbed the practice from Phoenician merchants. Cockpits were once as common in British villages as they are in the Philippines today. King Henry VIII built the Cockpit-in-Court in his main residence at Whitehall Palace in London in the sixteenth century, an era when most manor houses had their own pits, trainers, and feeding and fighting strategies for hundreds and even thousands of birds. James I was a cocker fan. And as William Shakespeare knew, the Globe Theatre was first and foremost a battleground for chickens rather than a stage for actors. "Can this cock-pit hold the vasty fields of France?" asks the chorus in the prologue to *Henry V.* The cheap seats in the Globe, called the pit, were where the roosters fought. Occasionally, many birds were let loose in a pit until only one was left alive, the origins of the phrase *battle royal.* The diarist Samuel Pepys, who went to a London match in 1663, spotted a parliamentarian betting with bakers and brewers, "and all of these fellows one with another in swearing, cursing, and betting."

Pepys felt sorry for the roosters and appalled by the losses incurred by poor men, but others were inspired. One early-seventeenth-century book asserts that cocking made men more courageous, loving, and industrious. "It is wonderful to see the courage of these little creatures, who always hold fighting on till one of them drops, and dies on the spot," the author Daniel Defoe wrote in 1724. A Scottish writer in that era proposed a "cock war" that would substitute conflict among humans for a match between royal courts, ending Europe's bloody struggles. At Newcastle upon Tyne in the 1780s, as many as a thousand roosters died during a week of matches.

William Hogarth's 1759 engraving, ironically inscribed *Royal Sport,* shows a chaotic London cockpit with a blind lord, pickpockets, and a hodgepodge of riffraff watching two roosters circle each other as a scandalized French gentleman surveys the savage scene. Cockfighting bans in the United States, Europe, and other industrial nations today focus on the cruelty of forcing an animal to fight to the death. Hogarth's scene reveals other concerns. The issue exposed in his drawing is not animal cruelty but human folly, dissolution, and greed. When Parliament banned cockfights in London in 1833, the reason was not to protect the birds but to halt criminal and disorderly behavior.

Those British parliamentarians, like American missionaries in the early-twentieth-century Philippines and today's Indonesian legislators, feared large groups of men from different social groups gathering to drink and gamble for hours on end. In rural areas such behavior might attract little notice, but in a rapidly industrializing economy it was seen as unproductive and potentially subversive. In the new cities—nineteenth-century London, twentieth-century Manila, twenty-first-century Jakarta—the poor are segregated from the rich, factory life demands promptness, and gambling is a sign of moral weakness.

"The old story about the rise of animal rights is a heroic and straightforward tale about Victorian Brits coming to their senses after countless millennia of slavery and animal cruelty," says Robert Boddice, a young British historian who has studied the movement. It wasn't that cockers were cruel to animals, but that such people lacked refinement. In the industrial era, gentlemen were not to mix with the lower classes, and working men should be by definition working. Upper- and middle-class women were often at the forefront of cockfighting bans in nineteenth-century Britain and America, as they were in the push for prohibition. Gambling and drinking often meant poverty and abuse for the women left at home, and prevented advance up the new social ladder.

The British Parliament banned cockfighting across England and Wales in 1835, and two years later Princess Victoria became a patron of the Society for the Prevention of Cruelty to Animals. As queen, she put the *royal* into the organization's name in 1840. Cockfights continued, but they moved to more discreet locations frequented mainly by working-class enthusiasts, and most aristocrats abandoned the sport. Though cockfighting lost its status and its legality long ago, foxhunting was permitted until 2005.

Americans were slower to give up the legal cockfight. It rivaled horse racing in its popularity, particularly in the South. After dining in Williamsburg with the Virginia royal governor in 1752 to discuss military matters, George Washington watched what he called a "great main of cocks" in nearby Yorktown. That same year, the capital's College of William & Mary banned its students from attending such events, a sign of their attraction. In recognition of the sport's popularity, modern-day Colonial Williamsburg keeps two gamecocks—Hankie Dean and Lucifer—though they are not allowed under Virginia law to fight.

The Virginia General Assembly had made cockfighting illegal in

1740, though cockers continued to import game fowl from abroad. Georgia followed suit in 1775 and the Continental Congress inveighed against the sport. In the new United States, cockfighting began to be viewed as a remnant of British barbarity, but its popularity persisted. There was a fleeting moment when a gamecock might have been part of the official seal of the new country. In 1782, a twenty-eight-year-old artist named William Barton included a game fowl on the top of his proposed seal. In the end, however, Congress chose a design featuring an eagle.

Thomas Jefferson avoided cockfighting and horse racing, according to accounts by his slaves. The military hero and Tennessee politician Andrew Jackson, however, was an enthusiastic gambler and cocker, and his presidential opponents used that against him. "His passions are terrible," Jefferson told Daniel Webster in 1824 as Jackson was vying for the presidency. Jackson insisted during his campaign that he had turned over a new leaf and had not attended a cockfight in thirteen years. Abraham Lincoln was reluctant to oppose the sport. He is quoted as saying that "as long as the Almighty permitted intelligent men, created in his image and likeness, to fight in public and kill each other while the world looks on approvingly, it's not for me to deprive the chickens of the same privilege."

Enthusiasm for cockfighting continued to wane during the nineteenth century. Mark Twain watched a match and marveled that spectators "lost themselves in frenzies of delight." He called it "an inhuman sort of entertainment" but added that "it seems a much more respectable and far less cruel sport than fox-hunting—for the cocks like it; they experience as well as confer enjoyment; which is not the fox's case." Women's groups attacked it along with liquor and other forms of gambling, and states began to pass laws banning its practice. By the 1920s, cockfighting was associated with gangsters and rumrunners, though it remained widely practiced in rural states like Oklahoma. The media magnate William Randolph Hearst, who pioneered the sports page, campaigned successfully to ban it in California.

Not until 2008, when Louisiana passed a law against it, did the last legal cockfight in the United States take place. In most states today it is a felony, although it remains popular in Appalachia and Hispanic communities. In Tennessee's Cocke County, for example, on the mountainous border with North Carolina, the sport remains commonplace if covert. Two large cockpit operations were raided in recent years as part of wider corruption investigations by the state. Thomas Farrow, who led one of the key sting operations, told a reporter that "it was like setting up a counterintelligence operation in the heart of Soviet Russia during the Cold War. There were no friendlies," he said. "During the day, everyone goes to church and goes to work, and it's like a Norman Rockwell community. But when the sun goes down, they all turn into vampires."

In 2013, a Tennessee state Republican lawmaker named Jon Lundberg tried to strengthen the state's cockfighting laws that currently make it a misdemeanor subject to a fifty-dollar fine. His attempts to put it on a par with dogfighting—a felony that involves a fine and jail time—failed. Shortly after the defeat, when I talk to him in his Nashville office near the state capitol, he is upset with opponents who argued that cockfighting is a Tennessee tradition that was enjoyed by Andrew Jackson and therefore shouldn't be demonized. "Slavery also is a Tennessee tradition that was upheld by Andrew Jackson," he notes drily. The lawmaker worries that the lax laws make his state a magnet for organized crime. "We've created an economic draw, but I don't think this is something that we want to draw people to Tennessee."

More than a century and a half after Jackson was criticized for his "passions," cockfighting briefly was at the center of a bitter 2014 primary fight between U.S. Senate minority leader Mitch McConnell (R-KY) and his Tea Party opponent Matt Bevin. McConnell supported a farm bill that increased penalties for participating in the sport, a move that the president of the United Gamefowl Breeders Association predicted would "destroy" McConnell's political future. Bevin attended a pro-cockfighting rally a few months later and said that "criminalizing behavior, if it's part of the heritage of this state, is in my opinion a bad idea, and I will not support it." His attendance and comments raised a national outcry, which intensified when a video of a campaign talk surfaced in which he describes tying the feet of chickens and swinging them over his head as a child prior to chopping off their heads. "Sometimes a foot would pop off," he added in a jocular tone, to the nervous laughter of the crowd.

In Manila, when I ask Luzong about New World cockfighting, he shakes his head sadly and confirms that the sport in Mexico, Colombia, and the United States increasingly is tied to the drug trade and crime syndicates, further blackening its already poor reputation around the globe. Like any blood sport, cockfighting has always been associated with men who like to gamble and drink, and as long as there have been cockfights, there have been poisons, magical spells, and any number of dirty tricks employed to beat rivals. But today, high-tech drugs and big bucks are the rule.

On the outskirts of a slum in Caracas, a gamecock breeder named Lorenzo Fragiel tells me that an unsettling change is taking place in the Venezuelan capital, which has a dozen rings within the city limits. "There's no longer a brotherhood of breeders," says Fragiel, a lean mechanic with a bristly black mustache. "People used to keep tabs on one another—it used to be friendly." The matches are more violent outside the ring. "There can be fights and broken glass," he adds with a shake of his head. "There's a lot of drinking." Ring owners charge an entrance fee and cock owners must also pay a fee per match and a fee for the referee, but the real money is in selling beer.

Fragiel keeps thirty-five fighting cocks in a spacious coop at one end of his little farm, each with a roomy cage. There is a good breeze, the floor is covered in clean sawdust, and there is no trace of the burning ammonia smell typical of most mass chicken operations. Some of the birds have livery like red jungle fowl. One variety called Bolos lacks a tail, and another—Papujos—has small feathers growing out of its cheeks. Both traits protect a bird in a fight. He has been breeding and fighting cocks since his godfather gave him his first bird at age twelve, three decades ago. "It's not a moneymaking operation," he says with a smile. "This is a passion."

Once he selects a potential new fighter, the bird acclimates to its surroundings for a few months. He cuts off the bird's comb so that a rival can't grab it in the ring, shaves it so that it doesn't suffer in the tropical heat, and removes the pointy spur. When the scab has healed and the bird is about a year old, training begins in a makeshift ring in the farmyard. He gives his birds only vitamins along with the feed his mother makes every week that contains his own recipe. "Each breeder has his own diet," he says. And he often will mix up a special feed for a bird that seems to be listless or ailing.

Fragiel uses a fighting-cock puppet to test out the bird's abilities. Later he will put it in the ring with another bird, though first he carefully wraps the beaks of both and attaches small plugs, or boots, over their sharp artificial spurs to avoid injury. Most cocks are not mature enough to fight until they are at least fourteen months old. The Venezuelan tradition is to use an artificial short spur made of tortoiseshell, a stark contrast to the wicked steel blades employed in other countries. Because tortoises today are a protected species, they use a plastic spur. He grabs one from a shelf and hands it to me. I'm surprised by how thin and light and blunt it seems compared to the lethal long metal spurs I saw in Manila. "There are so many Columbian breeders here now, that you now see more and more long spurs used," Fragiel says with a sigh. Those matches are aggressive and bloody, he adds. Of course, matches with long spurs are also usually quick affairs that last a couple of minutes. The Venezuelan fights can go on for fifteen or twenty minutes, since it takes longer for one bird to kill another.

He pulls me over to a cage in one corner, where he is nursing back to health a black-and-white bird injured in a fight. "The career of a fighting cock can end in only one way," he says. "Death." Fragiel's son walks into the coop, a twelve-year-old boy clearly at home with the cacophony of crowing. "If he is interested, that would be fine," the breeder says with a glance toward him. "But I won't push this on anyone—it has to be their own passion."

Cockfighting's decline in the nineteenth century began in Britain as the chicken took on the new role of feeding the urban masses. Yet it won't disappear in the next century, and it certainly remains vibrant in the Philippines, Venezuela, and the backwoods of Kentucky and Tennessee. It does face a long, slow fade in most countries with the rise of cities, the growth of alternatives like video games, and the growing awareness of and intolerance for animal cruelty. Plus, in our urban world, live chickens simply are not as visible and available. It is cleaner and simpler to let robots fight.

The ghosts of the ancient sport, however, remain hidden in plain sight in our common words and phrases. We might not be cut out for a job—a reference to a game fowl's feathers being trimmed—but we can still fight a battle royal, show pluck, remain cocky or cocksure, and sometimes have a set to. Whatever the moral or legal status of cockfighting, we will continue to guide our ships and planes from the cockpit.

6.

Giants upon the Scene

This blessed day will I go and seek out the Shanghai, and buy him, if I have to clap another mortgage on my land.

—Herman Melville, "Cock-a-Doodle-Doo!"

Order your chicken in simple white, basic black, or in nearly any color combination. Pick your size. A Serama bantam might be only eight inches high and weigh less than twelve ounces. The Jersey Giant can top twenty pounds. Malays that can stand on your floor and eat from your dining room table can be shipped overnight. Choose your disposition, whether ornery gamecock or calm Buff Orpington. If you decide you want a good layer or fat broiler or a general all-purpose bird, then click on, respectively, White Leghorn, Cornish Cross, or Barred Rock.

Such variety is a new phenomenon. Few Westerners took Marco Polo seriously when he described thirteenth-century Chinese chickens with hair like cats that lay large and tasty eggs. As recently as 1800, British ornithologists could count only five varieties on their entire island, and most were small, scrawny, and ill-tempered. By then, chickens had spread from their original home in South Asia to almost every corner of the world, from Africa's Cape of Good Hope to Alaska's Bering Strait. But the chicken was still a local bird. The fowl of the East and West had yet to meet to make the chicken of tomorrow.

Then, in 1842, Captain Edward Belcher, an accomplished explorer, naturalist, navigator, war hero, and one of the British Royal Navy's most detested men, returned to England after circumnavigating the globe.

"Perhaps no officer of equal ability has ever succeeded in inspiring so much personal dislike," concludes the *Oxford Dictionary of National Biography*, a source not given to exaggeration. The *Dictionary of Canadian Biography*, also a reliably staid reference, says that Belcher "suffered from an irritable, quarrelsome, and hypercritical nature which made relations with superiors and subordinates alike extremely difficult." Sailors repeatedly accused him of harsh treatment, officers did their best to avoid serving with him, and his young wife publicly charged him with knowingly infecting her with venereal disease on their wedding night.

Belcher was from an old Boston family—his great-grandfather was a colonial governor of Massachusetts also considered ornery and vindictive—that had fled to Canada during the American Revolution. In one of those odd quirks of history, the Belcher coat of arms may have served as a template for the United States seal that won out over the version featuring a chicken. As the Napoleonic wars raged, Belcher joined the Royal Navy at age thirteen. While still in his twenties he spent four years at sea, charting the waters from Alaska to Africa as a rising star in the officer corps. In 1831, Belcher was arraigned before a naval court for sending officers out on patrol without water and threatening to shoot a midshipman in the head for not obeying a signal. The judges acquitted him.

Shortly after, Diana Belcher brought the case against her husband to civil court. A fellow officer insisted that Belcher knew he was infected when he married, and a doctor who examined her noted that she also had been beaten. A judge dismissed the wife's request for separation, but she refused to live with him again. Diana ultimately got her revenge by modeling the wicked Captain Bligh on her despised husband in her bestselling book *The Mutineers of the Bounty*.

Though legally cleared of all charges, Belcher found himself with few friends and allies in the navy's clubby world. He was sent to survey the Irish Sea, a backwater posting at a time when the British Empire was rapidly expanding. His luck changed in 1836, when the captain of the HMS *Sulphur* sickened in Chile while on an extended mission to survey the Pacific Ocean. Belcher used his remaining connections to win appointment to the post. That October, just as a young Charles Darwin arrived back in England after his five-year voyage around the world on the HMS *Beagle*, Belcher crossed the Atlantic to take charge of the 380-ton and 109-man *Sulphur*.

Crossing the Isthmus of Panama to rendezvous with the ship, he stopped frequently to collect plants, animals, and minerals, including a twelve-foot alligator. "He was much attached to certain departments of Natural History," an officer later recalled. His cabin was like a museum. Belcher landed on the Galápagos Islands three years after Darwin, and was amazed by the ecological diversity. The captain ultimately contributed more than fifty specimens to the British Museum, the world's first national public museum. A British zoologist named a venomous sea snake that Belcher identified on the voyage after the captain—an inside joke, perhaps. It is the world's deadliest reptile. Darwin later claimed to have first identified the animal, but withdrew that assertion when given a copy of Belcher's narrative of the *Sulphur*'s voyage. Among the captain's donations to the museum was a male red jungle fowl.

Belcher's surveying mission of the Pacific coincided with the start of the first Opium War against China, then the most powerful, wealthy, and populous nation on earth. Westerners were eager to purchase the kingdom's fine manufactured goods like silk, porcelain, and tea, but the Qing court restricted their business to seasonal trade in the southern port of Canton, today's Guangzhou. Chinese authorities insisted on payment in silver, but British merchants began to substitute opium grown in their Indian domains. Not to be outdone by their rivals, American businessmen began to ship cheaper opium to China from Turkey. In the late 1830s, prices for the dangerously addictive drug plummeted, enticing millions of Chinese to take up the habit.

As the *Sulphur* charted Pacific waters in 1839, the emperor in Beijing had Canton harbor sealed and thousands of chests of opium seized and destroyed. Within months, British marines were on Chinese soil. Belcher arrived in the region in the midst of the hostilities. In January 1841, two ships under his command sank a Chinese fleet in the Pearl River Delta off Canton. Two weeks later he claimed the sleepy fishing village of Hong Kong for the newly crowned Queen Victoria, and the protected deepwater port quickly became a critical naval base for prosecuting the war. The captain was subsequently part of the British delegation that negotiated the surrender under the city's high walls. In August 1842, both sides signed a treaty that forced China to open its southern ports to Westerners.

That same month, Belcher arrived back in Britain. After an absence of six years, he was forty-three years old, with thinning red hair and windburned cheeks. Despite his exploits, he would have been unwelcome in the many London drawing rooms sympathetic to his abused wife, but he did have rare animal specimens to bolster his sorry social status. He carried with him a royal gift that would alter the trajectory of the industrial world.

Exotic animals from distant lands delighted the twenty-three-year-old Queen Victoria. She enjoyed what she called "the stream of barbaric offerings" like lions, tigers, and leopards that flowed from "tropical princes." Most of these she bestowed on the London Zoo, in Regent's Park, which by then had one of Europe's most extensive collections, reflecting the wealth and power of the growing British Empire.

She paid regular visits to the zoo. "The Orang Outang is too wonderful preparing and drinking his tea, doing everything by word and command," the monarch wrote in her diary that May after one excursion. "He is frightful and painfully and disagreeably human." The comment foreshadowed the controversy about evolution that would soon erupt. Darwin, who had visited the same ape, published a book on coral reefs that month, but secretly was working on a draft outlining his radical ideas about species development that he kept under lock and key.

The queen loved the circus, too, particularly the act where a handsome American who called himself "The Lion King" whipped a beast into submission, put his head in its open jaws, and then had it lick his shiny boots. Victoria attended the show several times when it played in London. After one performance she went backstage to watch the ravenous animals feed. Domesticated animals were to be pitied but wild ones were to be subjugated.

Victoria and her young husband, Prince Albert, preferred Windsor Castle, twenty miles west of the capital, to drafty Buckingham Palace, where the surly staff did little by word and command and construction made living unpleasant. The year before, she ordered a kennel built at Windsor to house her growing number of collies, terriers, and greyhounds. Thanks to suggestions made by her clever spouse, the canines enjoyed freshly cooked food, automatically refilled water bowls, and steam heat in their new accommodations.

In late September 1842, distractions were particularly welcome. She and Albert had just returned from their first trip to Scotland and Victoria was pregnant for the third time. She dreaded the resulting nausea and the months of confinement to come. The German prince was still viewed with great suspicion by many of her British subjects. Albert, wary of her beloved governess's influence, had convinced her to send the woman into exile. She was also under pressure from the current government to halt her correspondence with Lord Melbourne, a father figure and former prime minister opposed to many of his successor's policies. Meanwhile, striking workers threatened to bring the new industrial economy to a standstill. And three times that summer would-be assassins had attempted to shoot her.

The five hens and two cocks that rolled up in a carriage at Windsor on a fall day astonished and delighted the royal couple. These were not the typical short and bony chickens they were familiar with. The birds were quickly dubbed "ostrich fowl." The long yellow legs were bare and the feathers were a rich, glossy brown descending to an elegant black tail. They had black eyes and carried their heads held back against their bodies. The animals were quiet and calm and regal in their demeanor, and had little in common with the scrappy and ornery British birds like the Dorking likely brought by the Romans two millennia before.

They were called, confusingly, Vietnamese Shanghai fowl or Cochin China fowl but may have been from Malaysia. Western knowledge of East Asian geography in that era was hazy. Captain Belcher did not accompany the gifts; his scandalous past may have made it impossible for him to meet the queen. Nor did he leave written evidence for the birds' heritage. He had noted in his log the purchase of chickens at the northern tip of the Southeast Asian island of Sumatra, but he had also cruised the coast of Vietnam and spent time in southern China. Whatever their source, Belcher's exotic birds formed the kernel of a flock that would transform the chicken from a minor figure in the world's barnyard into its most important.

Victoria and Albert immediately decided to construct a new aviary to house the strange birds, replacing the small and decaying structure built by her grandfather King George III, the monarch who had provoked Bostonians and led Belcher's loyalist family to flee Massachusetts. Raised on a German country estate with aviaries, the prince had brought ornamental birds with him to England in 1840 when he married his first cousin Victoria. In December, the government approved 520 pounds for a new poultry-and-dairy farm at Windsor.

The couple chose a site in Home Park, a 655-acre private retreat east of the ancient castle where King Edward III had hunted deer and Oliver Cromwell had drilled troops. The new aviary would be close to the queen's kennels and within an easy stroll to Frogmore House, the estate that Victoria the previous year had given to her overbearing mother, the Duchess of Kent. As ranger of both the Home Park and adjacent Great Park, Albert oversaw the design and construction of the structure to house the queen's gifts and other birds. "Partly forester, partly builder, partly farmer and partly gardener," is how he once described himself. He had a practical interest in modern farming that dovetailed neatly with Victoria's love of animals.

The royal couple followed the project's progress closely. "We walked down to the Farm, after breakfast, to see where the new Aviary is to be," Victoria wrote in her diary in late January 1843. By the end of that year, it was nearly complete. The fanciful structure was a semi-Gothic affair of elaborate gables and finials topped with a hexagonal dovecote lined with mirrors "in which pigeons delight to gaze, and before which they are constantly pruning and dressing themselves," one visitor reported. The building, surmounted by a large weathervane topped with a rooster, consisted of a central chamber with two wings dedicated to roosting, breeding, and laying. It was a startling contrast to the small, dark coops typical of the day.

When the *Illustrated London News* gave readers a glimpse of the new aviary just before Christmas, the reporter described the design as showing "commendable regard to the conveniences of their graminivorous tenants." Spacious, dry, and warm roosts made of twigs of heather, hawthorn brambles, and white lichen emulated "their original jungles" and discouraged parasites. Heated flues running the length of the building provided warmth. Outside, a tall wire fence provided several runs for the birds during the day. Shrubs gave shade, and a veranda protected the valuable birds from chilling rain.

The complex included feeding stations, laying sheds, winter housing, and a veterinary clinic as well as several yards and small fields, all interspersed with lawns and gravel paths. Stately elms sheltered the yards from winter winds. "In cultivating the homely recreations of a farm, her Majesty has exhibited great industry and much good taste," the reporter concluded. The morning the article was published, Victoria and Albert were again at the complex admiring "some of the Poultry, which are really very fine & fat," the queen noted.

The same week, Londoners were busy snatching up copies of Charles Dickens's new novella *A Christmas Carol*, written to underscore the terrible conditions faced by London's working class amid the Industrial Revolution. In the story, even the hard-pressed Cratchits manage to buy a small goose for the traditional centerpiece of the English holiday meal. The fowl of choice that the reformed Scrooge orders from the poulterer is the prize turkey. But it was chicken rather than goose or turkey that was the talk of the 1843 Christmas table at Windsor. Guests were treated to "Cochin China pullets, which had been reared and fattened at the royal aviary in the Home Park," according to a newspaper account. "The Cochin China pullets weighed between six and seven pounds," two or three times the weight of a typical English chicken.

Soon the exotic fowls and their progeny were distributed to other royals around Europe. The following spring, after Victoria and Albert's uncle King Leopold of Belgium visited the aviary, eggs from the "more rare and curious" fowl from Asia were packaged and shipped to Brussels. By the summer of 1844, Victoria wrote that the complex was "really beautiful now, much enlarged with a terrace & fountain & a room is being built for us." In the small apartment, they could take tea and watch the birds with a view of the splashing fountain. They hosted the French king Louis-Philippe, whose favorite dish was chicken and rice, at the aviary that fall. Mainly, however, this was a private place for the couple to spend time "feeding our very tame poultry," as Victoria noted in her diary.

The young couple's fascination with fowl began to attract public attention and encourage imitation. The author of the 1844 book *Farming for Ladies; Or, a Guide to the Poultry-Yard, the Dairy and Piggery* saw the aviary firsthand, and concluded that the royal interest was for the moment purely amusement. But the writer adds that the long-term goal was to establish a poultry yard to supply the royal kitchens on a regular basis. "Nor can there be a doubt that the introduction of foreign breeds will thus—under the example of her gracious patronage—in the course of time, cause much improvement in the stock of our native species." Chicken, after all, was too expensive for most commoners, states the writer. "In London, the common prices of poultry are generally so high, that people of narrow income, if living in town, can seldom afford to put any on their table."

Chickens were already in Britain when Caesar arrived with his troops in 55 BC. The general noted the peculiar fact that it was against the law to eat them. Instead, he observed, the natives raised them "for their own amusement and pleasure," which suggested that they were used for both religious and gambling purposes. The Romans brought their own varieties of fowl across the English Channel, including one with an extra toe that is today called the Dorking, for roasting as well as for augury, cockfighting, and ritual sacrifice. Roman men carried the right foot of a chicken for good luck and ate rooster testicles to enhance their virility.

The oldest handwritten documents in Great Britain are the Vindolanda tablets found near Hadrian's Wall, which two centuries after Caesar's invasion separated Roman Britain from the Celtic Picts in what is now Scotland. Among these wooden boards is a shopping list given by a commanding officer of a garrison to his slave. The man was sent to a local market to purchase twenty chickens and "if you can find nice ones, a hundred or two hundred eggs, if they are for sale there at a fair price." The Latin saying "from the egg to the apples"—equivalent to our English "from soup to nuts"—described the Roman passion for starting a meal with an egg dish, and Roman bakers were perhaps the first to make custards and cakes using the hen ovum. Where there were Romans, there were chickens. And in Roman Britain, some people were even buried with the bird.

Chicken-and-egg eating declined after the Romans abandoned the island in the fifth century AD, but rebounded with the growth of monasteries in the early Middle Ages. The sixth-century AD Rule of Benedict forbids monks from eating the meat of four-legged animals, so fowl and their eggs emerged as an important commodity. Ducks and geese grew larger and fatter and made bigger eggs, but about half of medieval England's poultry were chickens. Most manor houses and monasteries kept at least a small flock. In times of famine from poor harvests or cattle disease, the chicken was the handy backup. It was, the food scholar C. Anne Wilson says, "the poor man's bird in medieval times."

Until Shakespeare's day, chicken was cheap. You could buy a whole bird for three or four pence, a pittance even then. It was, however, just one of many options. There were quail, bustards, herons, finches, wood pigeons, gulls, egrets, thrushes, mallards, and snipes. In thirteenth-century London, you could walk into a cookshop, put down eighteen pence—still a modest sum—and walk out with an entire roast heron. But as the human population grew and farmland expanded, wild birds became scarce and farmers began to raise large herds of pigs and cattle. Goose, pork, and beef became the flesh of choice. One historian calculates that, by 1400, chicken accounted for only 10 percent of what people spent on meat in eastern England. "Whoever could afford, substituted chickens with either goose or red meat," writes the Yale historian Philip Slavin. By this time, he adds, chickens were "a relatively minor contribution" to the late-medieval farm economy.

Swans and pheasant, by contrast, were strictly reserved for royalty. At Henry VI's 1429 coronation dinner in London, peacock was served as well. There were, fortunately, several other courses, since the colorful bird is infamously inedible. (The trend continued among the elite, and savvy Victorian diners knew to avoid the splendid peacock on the buffet.) Turkeys from the New World and guinea fowl from Africa arrived by the sixteenth century. In the early seventeenth century, doves were the rage, since they contributed to national security as well as to dinner tables. Besides providing delicious meat, their nitrogen-rich droppings extracted from dovecotes were an essential element in the manufacture of the gunpowder required for Britain's growing navy and army. Chickens and their eggs remained low on the culinary totem pole. Their profits, sniffed one agricultural handbook from the early 1800s, were "too inconsiderable to enter into the calculations of the farmer." Real farmers—that is, men—raised sheep and cattle.

In 1801, Parliament passed an act encouraging landowners to fence their land. The push for privatization had the effect of raising rents. In the decades that followed, poor and landless farmers decamped to the growing cities and burgeoning factories. In 1825, London's population of 1.35 million surpassed that of Beijing, which for more than a century was the world's largest metropolis. Food prices edged up. The British cleric and statistician Thomas Robert Malthus warned that disaster loomed. Less food and more people meant that "the poor consequently must live much worse, and many of them be reduced to severe distress." Ultimately, he argued, the only way the population would stay within the limits of the food supply would be through a combination of starvation, war, disease, birth control, and celibacy.

Today, demographers know that, as societies industrialize, birth rates spike as people leave the land for cities. But as incomes increase and women marry later and gain access to education, those birth rates steadily decline. Public health measures decrease infant mortality and infectious disease, technological innovations increase the food supply, and new transportation systems distribute that food more widely. This was as true for Victorian England as it was for early-twentieth-century America and twenty-first-century China. But in Victorian England, no one knew this. Malthus's grim warning appeared to be coming true as millions crammed into slums rife with disease and hunger, conditions that Charles Dickens was busy chronicling. Revolution, chaos, or apocalypse seemed nigh.

As Captain Belcher arrived back in Britain in the summer of 1842, a half-million workers in Britain were on strike protesting wage cuts as prices for basic staples rose. Supplying adequate and affordable food was essential to stave off revolution and keep Britain's new factories churning out finished goods to sell across its empire. Prince Albert emerged as a leading advocate of improving the country's dire food situation through the use of new technology. One of his tutors, Adolphe Quetelet, had been a colleague of Malthus's, and Albert was the royal patron of the statistical society that Malthus and Quetelet helped found. The prince encouraged the queen's interest in animals that might have practical benefits for his adopted country. "Agriculture needs encouragement from the Crown," he writes in a condescending note to his father, "which V. naturally cannot give."

The transformation of British agriculture was already apparent at Leadenhall Market, a huge stone building that sat on top of the Roman basilica and forum of ancient Londinium, where vendors two millennia earlier hawked live chickens. Today, it is an upscale mall, but in the fourteenth century it emerged as the center of the country's poultry industry. Gaggles of honking geese were herded through the city's streets from outlying farms and ships docked at the nearby Thames to unload French eggs. By the 1840s Leadenhall was the world's single largest poultry market, selling two-thirds of the city's fowl, including geese, ducks, turkeys, pigeons, and game fowl, as well as chickens. But the market's famous chaos, filth, and noise were largely gone. Thanks to the new railroads—Victoria and Albert took their first train trip a few months before their exotic chickens arrived—the animals could be slaughtered on a distant farm and then brought to the capital while still fresh. It was a radical innovation, given the difficulty and expense involved in shipping, storing, and tending to live ones. Refrigeration still lay in the future, but this change marked a big step toward the plastic-wrapped meats of today's supermarkets.

British farms could not begin to supply the 4 million birds that Leadenhall and other London markets sold each year in that era, or the enormous number of eggs demanded by Londoners. Most poultry was brought by ship from Ireland, Holland, Belgium, and particularly France, where the industry had long been more highly developed. The "three French hens" bestowed in "The Twelve Days of Christmas" is a reminder of the age-old English dependence on French poultry. As Britain's population soared, so did the imports. Britons ate about 60 million foreign eggs in 1830 and more than 90 million by 1842. Eggs were also used to soften leather; one factory alone bought eighty thousand eggs a year to produce kid gloves. The British reliance on foreign nations for such a basic foodstuff worried Albert.

As the Windsor aviary neared completion in 1843, the royal couple appointed a full-time royal poultry keeper named James Walter, praised by the *Illustrated London News* as the chickens' "vigilant guardian" who "understands their language, their dispositions, their diseases." A large drawing shows him covered in a dozen adoring pigeons, two of which are vying for a perch on his top hat. He immediately began to conduct breeding experiments to see if the Dorking chicken, the variety likely brought by Caesar's troops, could be crossed with the Asian ostrich fowl.

"In order to improve the breed of the genuine Dorking fowl, that it should be crossed with that of the Cochin China fowl, the necessary arrangements were made," the *Berkshire Chronicle* reports on September 28, 1844:

"A Dorking hen, which has been roosted for some time past with the fowls from China, has recently been in the habit of laying twice, and sometimes thrice a week, eggs containing double or two distinct yolks. Mr. Walters [*sic*], determined to try the experiment of attempting to hatch one of these double-yolk eggs, placed it, with several other eggs, under the hen. The result was that two chickens were produced from this single egg; one is a cock bird of the pure Cochin China breed, and the other is a hen chick of the Dorking species, both of which are now five days old and in good health."

The publicity surrounding the royal chickens began to attract widespread interest. London's first poultry show took place at the zoo in Regent's Park on June 14, 1845, three years after Victoria visited the orangutan. The modest affair was something of a novelty; the first major dog show in London wouldn't take place for more than three decades. Agricultural exhibitions were mostly designed for farmers rather than animal enthusiasts; pets were largely an upper-class luxury. Participants walked past the bear pit at the back of the garden, and there was no tent to shield exhibitors from the unusually damp weather afflicting Britain that season. There were only a dozen varieties of chicken, including some from Spain and the Madeiras, at least one said to be from China, and one from Malaysia. Most entrants were, however, domestic birds, and a speckled Dorking won first prize.

Europe's wet June continued into July, when a strange disease began to destroy the potato crop, the primary staple of the rural poor across the continent, in a province in King Leopold's Belgium. Wind blew the stench of the rotting plants with the destructive spores and soon it spread to Prussia and France. In September, it appeared on the Isle of Wight in southern England, where Victoria and Albert were vacationing at the royal residence at Osborne that they had just purchased, and where Victoria would die more than half a century later. "Another fine morning, when we walked down to the beach, where the Children were playing about happily," she writes September 13. That day, the potato blight was first reported in Ireland. It was the start of what would be remembered in Europe as the Hungry Forties.

Ireland was then ruled from London, and nearly half of its 8 million overwhelmingly rural people depended solely on the New World tuber for their primary sustenance. Life was already hard in a country where most Irish eked out a living as tenants on land owned primarily by wealthy English and Scottish landlords. An 1845 government report issued before the disastrous fall harvest said "it would be impossible to describe adequately the privations which they and their families habitually and patiently endure."

The commission, appointed by Prime Minister Sir Robert Peel, found "that in many districts their only food is the potato, their only beverage water" and that "a bed or a blanket is a rare luxury." One English lord warned the prime minister that the Irish peasant was supported solely by a quarter acre of potatoes. "Deprive him of this, or let his possession be uncertain, and what interest has he in preserving the peace of the country?"

At Windsor on November 6, Victoria and Albert walked to the aviary after breakfast and took a drive on an unusually sunny and warm day, welcome after the dreary summer. She was relieved to hear later from Peel that the situation in Ireland was not as bad as he had first feared. "Indeed, I think the alarm must have been exaggerated," Victoria writes in her diary. But while Holland, Sweden, Denmark, and Belgium acted quickly to import grain and regulate prices in their own lands, Parliament dithered. A month and a day later, a frustrated Albert sent a harsh memo to the government criticizing the inaction. "Half of the potatoes were ruined by the rot, and . . . no one could guarantee the remainder," he complains.

That winter, outraged Irish journalist John Mitchel, later convicted of treason and exiled, warned that a devastating famine had begun among his people as they watched "heavy-laden ships, freighted with the yellow corn their own hands have sown and reaped, spreading all sail for England." In London, those ships were seen as essential for the industrializing heartland. Without the Irish larder, factory workers would face food shortages that easily could explode into revolution. Meanwhile, the landed gentry opposed repealing laws that imposed steep tariffs on grain and other foods from abroad. And there was more than corn and grain flowing out of Ireland. In 1841, the Irish sent an estimated 150,000 cattle, 400,000 sheep, and 1 million live pigs to England and another 400,000 pigs that had been slaughtered and turned into pork and bacon.

An unknown number of chickens were also exported. "Eggs also constitute a by no means unimportant article of commerce," one contemporary British writer notes. "The largest supplies to the English markets are obtained from Ireland." In 1835, 72 million eggs were sent to England. That figure grew throughout the famine of the 1840s and, by the early 1850s, "it is estimated that we yearly receive one hundred and fifty millions. Of this number London and Liverpool respectively consume twenty-five millions each."

On February 23, 1846, the House of Lords bemoaned the state of Ireland, but the focus was on the country's lawlessness rather than the hunger at its root. Meanwhile, desperate people eating rotting potatoes suffered from terrible bowel disorders. Fever spread in County Cork. A relief commission warned that what was left of the noninfected potatoes would last only until April. The Royal Dublin Society, an organization created to keep the struggling island abreast of the latest agricultural, industrial, and scientific advances, decided to act by voting to offer a gold medal and twenty pounds for the best essay on the potato crop blight.

Then the society officers moved on to plan their large spring exhibition. Prize cattle and other domestic animals from across the United Kingdom were to be judged for size and beauty. Prince Albert, who had recently become a patron of the society, made plans with Walter to ship some of the exotic and crossbred birds across the Irish Sea to show at the exhibition in the Irish capital.

The potatoes didn't last until April. Food riots broke out by March. Lord Heytesbury, the British representative in Ireland, counseled calm, and grain, beef, and poultry continued to be exported from Irish harbors to English markets. On March 23, as Albert helped Walter design the cages to transport three of the crossbred hens and one cock from Windsor to Dublin, a man in Galway was reported to have died of starvation. A million more deaths were to come, while another million would flee the country, most destitute. Those survivors would clamber aboard what became known as coffin ships that transported the sick, starving, and suffocating people to Canada and the United States.

The first famine death went unnoticed in London, but the Cochin fowl's journey to Ireland made the news. A London paper reported on April 17 that the four exotic chickens had arrived in Dublin "perfectly free from injury." The landlords and well-to-do farmers who attended the show were in awe of the queen's entries presented by Walter on the Dublin fairgrounds. He explained to a rapt audience that one of the hens, Bessy, had laid 94 eggs in 103 days, an extraordinary record at a time when half that number would have been notable. There were skeptics. "But if this be a fact," said one wag of Walter's claim, then "there is no limit to the improvement, of which these double-barreled Hens are capable, till by the aid of forcing and extra diet, they become, like Mr. Perkins's steam gun, able to discharge Eggs at the rate of several dozens in a minute."

The birds "created such a sensation, from their immense size and weight, and the full, deep tone of the crowing of the cock, that everybody was desirous to possess the breed, and enormous prices were given for the eggs and chickens," according to the contemporary poultry aficionado Walter Dickson. "With respect to beauty," he adds, "they have certainly nothing to boast of." And those were the cocks. "The hens are still more ugly." Meanwhile, one exhibit-goer watched with horror as hungry locals crept into pens to steal what was left of raw turnips rejected by the well-fed prize cattle.

Walter presented the birds as gifts to Lord Heytesbury from the queen before he departed with three gold medals for Windsor. Heytesbury regifted two of the birds to the Dublin poultry breeder James Joseph Nolan. He was impressed by the birds' large eggs and tasty meat, and grew certain that such improved poultry could halt the mounting deaths across Ireland. Pigs were as dependent as humans on the potato plant, and they quickly starved or were slaughtered. Chickens, by contrast, could eat weeds and insects indigestible to humans who, in Ireland, had turned to consuming leaves and grass. Poultry was a vital defense against hunger, argued Nolan. Yet during the worst years of the blight, he watched in disgust as landowners continued to export a million pounds' worth of chickens and eggs to England. "Instead of persecuting the poor," the gentry should procure the new productive chicken breeds and supply their tenants with eggs, he wrote in 1850. These could quickly proliferate. Then the landowners might be "respected, beloved, and venerated, and the strength and sinews of the land would not be crossing the Atlantic."

By then it was too late, at least for Ireland. The slums of New York and Boston were already jammed with those who managed to escape the famine and survive the terrible voyage. Ireland's population fell by half and still has not returned to the levels of Victoria's early reign, a bleak testament to the long-lasting impact of a disaster that was as much political as biological. Nolan's vision of poultry as a savior for the rural poor was prescient, and was echoed by others in the decade that followed the Hungry Forties. What lay the foundation for the modern chicken was neither a contrite Irish gentry nor a remorseful British government, but a backyard chicken phenomenon known as "The Fancy" that makes today's movement pale in comparison.

In the decade between 1845 and 1855, Britain and America were gripped by an obsession with exotic chickens. The fad, like most economic bubbles, left many disillusioned and poorer, but it also aided Darwin in his attempts to explain evolution, transformed the lives of women, and led to the modern industrial chicken that feeds much of our species today.

"Events which are injurious while they take place, often leave good results behind them," notes Elizabeth Watts, the first female editor of a regular English-language publication. Watts was referring to the railway bubble that, when it burst, devastated thousands of British investors in the late 1840s. Even Darwin and the writer Charlotte

Brontë felt its sting when companies failed by the dozens. All that speculation did result in new rail lines that made traveling easier, she added. As editor of the first regular publication devoted to chickens—*The Poultry Chronicle*—Watts was drawing a parallel with what some at the time derided as "hen fever."

An unmarried and enterprising woman of means who lived in the fashionable London quarter of Hampstead, she was an early owner of Cochin China fowls and a player in the chicken mania. In 1854, Watts exchanged some of her Cochins for the Sultan breed of chicken from Istanbul. "They arrived in a steamer," she writes. "The voyage had been long and rough; and poor fowls so rolled over and glued into one mass had never been seen." The imports, said to have wandered in the tulip-studded gardens of the Ottoman sultan on the banks of the Bosporus, were an immediate hit in London. The snow-white feathers that reached to the toes and its spectacular puffball crown were a media sensation; Darwin later mentioned Watts's description of the bird admiringly.

What began as an eccentric royal hobby now was a national craze. "If you travel by a railway, poultry becomes the subject of conversation among your travelling companions; a crowing neighbour salutes you from another carriage, or perhaps a fine Cochin China is held up at the window to show the beauty of his points," reads one editorial in the *Chronicle*. "If you lose sight for years of some acquaintance, you are pretty sure to meet him at a poultry show, or to encounter his name among the exhibitors."

The 1846 fair in Dublin on the eve of the potato famine lit The Fancy's fuse. "The eggs having been freely distributed, with that gracious kindness for which Her Majesty and Prince Albert are so celebrated, the breed may now be easily obtained," one contemporary author noted. Sleek clipper ships brought new varieties to Britain such as Cochins or Shanghais purchased by British sailors in the recently opened port of Shanghai. Unlike the tall and ugly Windsor bird, this one boasted feathers on its short legs, a small, soft tail, and a wide body as puffy as a pillow. Some were black, others white, and still others a delicate buff color. Their feathers were like soft clouds.

They weighed, on average, twice that of an English chicken. Unlike the scrappy Dorking, these were calm and gentle.

Collectors eagerly awaited the arrival of the next clipper ship. In December 1848, chicken and pig enthusiasts from around the British Isles gathered in the industrial city of Birmingham. At the time, Watts's publication noted, keeping poultry was still widely regarded as "an idle whim from which no good result could by possibility accrue." Huge crowds packed the exhibition. "Cochins came like giants upon the scene; they were seen, and they conquered," writes one nineteenth-century poultry historian. "Every visitor went home to tell tales of the new fowl that were as big as ostriches, and roared like lions, while [they] were gentle as lambs."

And they were soon as expensive as precious metal. The next year, at the second Birmingham show, one vendor sold 120 birds for the equivalent of fifty thousand dollars today. A decade after Victoria built her royal henhouse, the *Times* reported that "poultry shows, and many of them upon a large scale, are being held now in all parts of the country." At London's popular summer Bartholomew Fair, buyers snapped up Cochin China fowl with posh names like "Marie Antoinette," "The Regent," and "Richelieu" for the equivalent of twenty-five hundred dollars apiece. "People really seemed going mad for Cochins," a contemporary writer noted. The satirical magazine *Punch* had a field day, publishing cartoons lampooning women walking their giant chickens on leashes.

An 1852 London exhibition under the patronage of Prince Albert drew more than five thousand people, and buyers had to beware of plumage tinting and hackle trimming that might disguise a plain bird as a more exotic creature. Eight detectives kept watch to prevent theft. There were, it was whispered, organized gangs of poultry swindlers at work. One woman paid the equivalent of two thousand dollars for a pair of Cochins. More than eighty thousand dollars changed hands in one day alone. The *Times* was disturbed by what it called "the seductions of this new mania" that "rages among us with epidemic fury." By 1855, the fever broke and prices plummeted. In August of that year, *The Poultry Chronicle* was quietly folded into *The Cottage Gardener*.

"They have fallen in price because they were unnaturally exalted," one of Watts's last articles states. But she confidently predicted that large and fast-laying birds "shall now be within the reach of all." Breeding the Asian birds with local varieties like the small and ornery Dorking soon gave chickens an advantage over the geese and ducks long favored in medieval England. They could eat a wider variety of food, were amenable to living in small spaces, and produced more eggs over a longer period of the year. They were also gentler than geese and provided much more meat than a dove. Even the stuffy *Times* saw the possibilities. "If the Cochin China breed will really give us poultry of a finer and cheaper description than we have had before," the paper stated, "the 'mania' will have done its proper work."

7.

The Harlequin's Sword

Amidst the immense number of different breeds of the
gallinaceous tribe, how shall we determine the original
stock? So many circumstances have operated, so many
accidents have occurred; the attention, and even the whim
of man have so much multiplied the varieties, that it appears
extremely difficult to trace them to their source.

  —Comte de Buffon, *The Natural History of Birds*

The bird curator at the Natural History Museum gives me a dubious look. "No one comes to see Darwin's chickens," Joanne Cooper
says. The ugly concrete annex attached to an old Victorian building
in Tring, a village north of London, is the Vatican of ornithology,
where a reluctant and eccentric banker named Baron Lionel Walter
Rothschild collected more than 2 million specimens and a hundred
thousand species of butterflies and birds before he died in 1937. Now
public, the museum houses three hundred thousand bird skins obtained primarily during the heyday of the British Empire as well as
thousands of skeletons and stuffed birds from around the world.

  Cooper, wearing a ponytail and glasses, guides me through security and down the bare corridors. Over her shoulder she tells me that

researchers interested in Darwin typically ask to see the finches the naturalist collected in the Galàpagos Islands during his HMS *Beagle* expedition or the pigeons that he later bred and studied as part of his effort to understand the workings of evolution. His 1859 book *On the Origin of Species* begins by discussing pigeons, and he dwells at length on the minute differences in the beaks of the South American finches.

Largely forgotten are the chicken carcasses that he donated to the museum a century and a half ago, as well as the story they tell of how the common fowl helped the naturalist clinch his controversial theory. In the midst of The Fancy and encouraged by one of its leading figures, Darwin sought specimens of the fowl from around the world, carefully studied their attributes, and waded into the heated debate about what made the chicken the chicken.

Modern chicken research is dominated by challenging but narrow goals as scientists try to eke out an ounce more meat from a pound of feed, prevent disease transmission in a warehouse crammed with tens of thousands of birds, or figure out an inexpensive method to determine if an egg will produce a hen or a rooster. In the mid-1800s, the field was crowded with theologians, philosophers, and amateur breeders as well as biologists. Chickens were at the heart of the bitter culture war pitting atheists against believers and abolitionists against slavery advocates.

The bird's modern identity took shape a century earlier in the Scandinavian town of Uppsala, where Carolus Linnaeus, a Swedish naturalist, was fascinated with the fowl. This central figure of the Enlightenment, a small-town university professor adored by Voltaire, Goethe, and Rousseau, kept a small flock in his large garden behind the modest two-story house that still stands on a corner on the old town's outskirts. Students crowded into his spacious and sunny study on the second floor to hear his lectures, and he occasionally discussed the bird's many medicinal and culinary properties, describing various Swedish breeds, recounting chicken folklore, and explaining how to neuter a cock.

When he created his system of taxonomy, which we still use today, he put the domestic chicken together with a wild variety said to live at the mouth of Southeast Asia's Mekong River—an early reference to what we now call the red jungle fowl.

In Linnaeus's animal kingdom, the chicken falls into the phylum Chordata, since it shares, with humans and other vertebrates, a backbone. The class is Aves, the collection of ten thousand species we call birds. Galliformes is their order, a group that includes a host of heftier species like turkeys that prefer the ground to the air. Their extended family is Phasianidae, which includes all pheasants and partridges and quail and peafowl, lumped together because they share a certain look that includes spurs on their legs, squat bodies, and short necks. Four sorts of jungle fowl fall into the more intimate subfamily Phasianinae, along with such charmingly named birds as the Tibetan-Eared, the Mikado, and Mrs. Hume's pheasant.

The genus *Gallus*—the Latin word for rooster—consists of four sibling species of jungle fowl. Red jungle fowl and chickens form a single distinct species, *Gallus gallus*, though some biologists give the barnyard bird the name *Gallus gallus domesticus*.

By putting a wild and domesticated bird together, Linnaeus was not suggesting that the chicken evolved from the wild one, only that they were similar enough that one could breed with the other. "God created, Linnaeus organized," he explained. Like most Europeans of his day, Jews and Christians alike, Linnaeus adhered to the belief that God made wild as well as domesticated plants and animals during the first week of creation, as recounted in the biblical book of Genesis. Species, therefore, did not change. Horses and donkeys might mate, for instance, but the resulting mule would never have its own progeny.

The French naturalist Georges-Louis Leclerc, also known as the Comte de Buffon, was one of the few in that era to challenge Genesis—and Linnaeus—publicly. Though both men were born in 1707, they lived in radically different worlds. The French count was a high-living Parisian aristocrat who wrote a natural history that scandalously suggested a much older earth than outlined in scripture. His connections and charm in prerevolutionary France kept him safe from the clerical authorities. Buffon did not directly challenge the idea of fixed species, but he thought that Linnaeus's system was too arbitrary and confining. Within a species, he pointed out that enormous variation was possible, citing "the Laplander, the Patagonian, the Hottentot, the European, the American, and the Negro," among humans.

The varieties of chickens, he noted, were even more pronounced. There was one Asian variety of poultry, Buffon had heard on good authority, with bones "as black as ebony." This is an early reference to the Hmong chicken found in south China and Vietnam that has not just black feathers but black meat and blood and is prized for its taste and medicinal properties. Buffon believed that such variation must be the result of human influence and not just the static creation of a supreme being. "The cock is one of the oldest companions of mankind, and . . . among the first which were drawn from the wilds of the forest, to become a partaker of the advantages of society," he asserted. Wild ones still existed in the forests of India, but he suggested that in many places "the domestic fowl has almost everywhere banished the wild one." The naturalist all but said that the domesticated bird evolved from a wild creature to a human companion, but didn't explain what mechanism could engender such a transformation.

Not until the year that Darwin was born, 1809, did the concept of change among species capture widespread attention. The French naturalist Jean-Baptiste Lamarck claimed that species could transmute into other species as part of a mysterious force that made life ever more complex. Environment, Lamarck insisted, determines traits: moles don't need eyes nor birds teeth. He argued that an individual animal—say, a giraffe that stretched to reach the leaves of a high tree—could pass on these traits to the next generation. Critics, both colleagues and clergy, pounced on the theory as either flawed or blasphemous. In a clever stunt that drew a lot of publicity, one distinguished French scientist, Georges Cuvier, closely examined animal mummies brought back from Egypt during Napoleon's attempt to conquer the ancient land along the Nile. The ibis and cats were no different from our modern varieties, he noted, incorrectly concluding that species were fixed and that Lamarck was grossly mistaken.

Yet evidence mounted that the world was far older and life far more varied than could be accounted for in the Bible, which, after all, doesn't specifically identify chickens in the Old Testament. A steady stream of living creatures that explorers, conquerors, and colonizers sent back to Europe would have required an unimaginably vast Noah's ark. The new science of biology grew up around these specimens. Yet as these discoveries challenged religious beliefs, they also were used to justify slavery.

In the late 1840s, just as dozens of strange and novel Asian fowl varieties spread across Europe and America, some scientists argued that the chicken's remarkable diversity demonstrated that its roots were in more than one ancestor and that humans might have different roots as well. The Philadelphia Quaker and physician Samuel George Morton maintained that chickens and guinea fowl could produce fertile offspring—a claim that later proved dubious—and that blacks and whites, therefore, could do the same and still be classified as separate species. Slavery's supporters in the American South quickly publicized his findings.

A plucky Southerner refuted this claim with his own poultry experiments. The Lutheran minister and amateur ornithologist John Bachman made the case for a single ancestor of all chickens. The Charleston pastor closely observed red jungle fowl in the London Zoo as hen fever raged in Britain. "So nearly did they resemble the domesticated fowl in every particular, that we could scarcely distinguish them from some varieties among our poultry," he wrote in 1849. His own breeding research on several species of jungle fowl demonstrated that two different species could not produce fertile offspring and concluded that human traits, like those of poultry, might vary considerably within the same species.

Darwin, who, in one of those interesting quirks of history, was born on the same day as Abraham Lincoln, closely followed this contentious debate in scientific proceedings, conferences, and journals. An ardent opponent of slavery since witnessing its awful consequences on his *Beagle* voyage, he was appalled that progressive scientific ideas were interpreted as an excuse for barbarity. Darwin's insatiable curiosity often overrode the class, race, and professional boundaries typical in his day. He once paid for lessons in taxidermy from an African living in Edinburgh, remembering him as "a very pleasant and intelligent man."

His open-mindedness extended to species that others considered inferior or unimportant. Darwin spent seven years examining barnacles. Listed in a contemporary directory as "farmer," he also knew that amateur breeders, whatever their social or financial status, were vital sources of information. So when a devout country rector of modest means but enormous expertise urged him to put away the crustaceans and focus on poultry, he eventually followed the advice. This led Darwin to the hard evidence he needed to make a convincing case for his theory of evolution.

Chickens may outnumber every other kind of bird today, but they get short shrift among ornithologists. "The domesticated specimens are not as valued," curator Cooper tells me as we pass two elderly women at a wooden table examining stuffed kingfishers from Sudan. "But for Darwin, it was the smaller domesticates that he needed to establish his fundamental principles." For four frenzied years, at the tail end of The Fancy, the naturalist focused his attention on pigeons and chickens. Small animals that breed quickly in captivity are good models for demonstrating the long-term power of natural selection.

Darwin worked mainly with skeletons rather than skins, Cooper says as we climb the stairwell to the next floor. Skeletons, she explains, are just that, but to prepare skins, "You remove their innards, pop out the brains and eyes, and sew them back up." We reach the annex floor devoted solely to skin storage, where I lose count of the identical narrow aisles made by double rows of off-white cabinets. The fluorescent lights illuminate a cheery yellow floor, yet the vast room has a sepulchral quality. Without hesitating, Cooper glides down one aisle and pulls open a drawer like those used for lateral hanging files.

Inside are a dozen birds lying in their shared metal coffin, each with a tag on a toe as in a morgue. The skins resemble deflated balloons. She reaches for one. "Collected by A. R. Wallace. 1862," reads the tag. Neat handwriting identifies it as *Gallus bankiva* from Malacca on the Malaysian peninsula. This is a red jungle fowl variety that the naturalist Alfred Russell Wallace encountered near the southernmost native range of the bird, between Singapore and Kuala Lumpur. Wallace independently came up with the theory of evolution while traveling in Southeast Asia, and his writings on the subject were read together with Darwin's at a London meeting in 1858.

In the drawer below are two stuffed red jungle fowl, fresh enough that they seem ready after a century to make a break for the forest. There are dull gray-brown hens and showy roosters lying on their backs, beaks pointed to heaven. The colors of the males in the monochrome surroundings are dazzling. There is a shower of gold around the neck, a magnificent patchwork of royal blue across the body, feathers with a subtle burnt orange brightening to yellow and violet turning to black at the tail. In another drawer, a 1939 specimen is labeled as "Domestic Guinea Fowl x Domestic Fowl." Another contains hefty chickens with naked necks. A sign on the outside of the drawer explains that they are a breed "periodically featured, incorrectly, in the media as a hybrid between chicken and turkey." Sometimes species are just species.

The Reverend Edmund Saul Dixon and Darwin inhabited eerily parallel worlds before their correspondence began as The Fancy gathered steam in 1848. Both men were on the verge of turning forty and had graduated seventeen years earlier from Christ's College at Cambridge University. And both men had contemplated a rural ministry as a way to pursue their intense interest in the natural world.

What set Darwin on a different path, aside from his family's wealth, was a surprise invitation to go on an around-the-world trip as interesting company for the captain of the *Beagle*. Shortly after Darwin set sail on what became his famous five-year voyage, Dixon was ordained. While Darwin roamed Patagonia and gathered his Galapàgos finches, Dixon married and moved to a quiet rural parish a hundred miles east of the capital in a modest house with a small backyard and bought a few chickens on a whim. Darwin returned to

England to settle with his own wife on an estate in the rolling Kent countryside south of London in September 1842, the same month that Captain Belcher bestowed his gift on Queen Victoria.

Dixon was puzzled by the contradictory advice in the few books on poultry rearing available at the time, and, after observing their behavior and breeding habits, became convinced that chickens "afford the best possible subjects for observing the transmission or interruption of hereditary forms and instincts." When his wife suddenly died, the widower embarked on an obsessive study of poultry and began to attract attention in the small world of naturalists for his expertise.

In 1844, an anonymous author in Britain published *Vestiges of the Natural History of Creation*, a slim volume that argued that everything, from planets to plants, developed from earlier forms, which its writer envisioned as an "organic process." Species evolve. It became a bestseller. Prince Albert read it to Queen Victoria, perhaps in their aviary retreat. But it also drew critics, like Dixon, who argued that the science was flawed and the conclusion blasphemous. He embarked on a series of experiments to demonstrate that different species could not produce fertile chicks, and therefore that species cannot change.

In a gossipy letter to his botanist friend J. D. Hooker in October 1848, Darwin related that their geologist acquaintance Charles Lyell had been knighted by Victoria and spent time at the queen's new Scottish lair of Balmoral. He also went on at length about his barnacle research. "Though I have done little about species," he mentioned in passing, "I have struck up a cordial correspondence with a first-rate man." He had asked Dixon if he would examine newly hatched chicks to see if they showed the same characteristics found in adult birds. Dixon, delighted to assist the esteemed scientist, readily agreed and encouraged Darwin to set aside barnacles and put more emphasis on poultry research.

Two months later, Dixon sent his newly completed poultry book to Down House with a personal inscription. A copy also went to another notable acquaintance, Charles Dickens. It was the end of a year of revolutions that had begun with a bang but ended with a whimper. Monarchies tottered, barricades went up, and Marx and Engels published *The Communist Manifesto*. Though it was unlikely to bring partisans into the street, Dixon's *Ornamental and Domestic Poultry: Their History and Management* was the English rector's own manifesto, published the same month that the Birmingham show—where he was a judge—set The Fancy into high gear.

On Christmas Day, as Victoria and Albert entertained around a decorated evergreen tree in Windsor, starting another trend, Darwin noted in his diary that he had finished a book on Siberia that was "dull" and another on natural history museums that he found "inclusive." He also mentioned reading Dixon's lengthy volume but withheld comment. A detailed compendium of poultry varieties, the book gave specific instructions on caring for and feeding the birds. It also was a forceful and erudite rejection of what the author called "the wild theory . . . so fashionable of late, that our tame breeds or varieties, are the result of cross breeding between undomesticated animals." Dixon lambasted the idea that species could change, based on his experiments with separate species like chickens and guinea fowl, which were, he wrote, incapable of producing fertile progeny. Each species was locked within its own cage of traits that could never be shared with another, and therefore, no new species were possible. Genesis was the beginning and end of creation. Chicken varieties were permanent and ancient, produced by a creative power rather than human intervention.

Dixon even cited his colleague in an attempt to bolster his case. "Mr. Darwin's discovery, the result of his great industry and experience, that 'the reproductive system seems far more sensitive to any changes in external conditions, than any other part of the living [system],' confirms my suspicion," he wrote. Reproductive systems, in others words, were the least adaptable animal part. Creating hybrids of two species with working reproductive systems was extremely improbable.

Darwin underscored Dixon's strong views on the immutability of species in the copy the author had sent. On a blank page at the volume's end, he mused that many domestic breeds might descend from wild birds, "yet I think others cannot probably have come from their crossing." Then he crossed out his own comment. It wasn't until the following March that he offered a catty review to a friend. "It is very good & amusing," he wrote slyly. What amused Darwin was Dixon's attempt to halt the public discussion of species change.

The two men continued their correspondence, exchanging extensive notes on Dixon's breeding experiments, though most of it is lost. The Down House naturalist, however, kept his distance, pleading illness. "I now really hope that your recovery may afford me the opportunity & privilege of making your personal acquaintance some time before the Winter arrives," Dixon wrote that spring. There is no record that the two men, so similar and yet of such different opinions, ever met. But Darwin soon began to consider how both chickens and pigeons might bolster his case for natural selection.

Dixon grew shrill in the years that followed and dismissed the idea of wild animals turning into domesticated beasts as "a sort of harlequin's sword" that could "convert a clown into a columbine." In 1851, he declared that "the Almighty gave to the human race tame creatures to serve and feed it," adding that a domestic fowl could not have derived from the red jungle fowl. Such a theory was not only wrong from a scientific point of view. It was also "a great heresy."

Darwin kept quiet publicly, unprepared to make his argument for species change through natural selection, but he groused to his trusted second cousin William Darwin Fox that Dixon "argued stoutly for every variety [of chicken] being an aboriginal creation" while he also "seemed to entirely disregard all the difficulties on the other side."

In 1855, as The Fancy bubble burst, Darwin set aside his barnacles and turned to collecting and examining poultry. He had a lot to learn, but was no longer in touch with the knowledgeable but acerbic Dixon. "I was so ignorant I did not even know there were 3 [varieties] of Dorking Fowl, nor how do they differ," he wrote his cousin that February. "It is an evil to me that Mr. Dixon is such an excommunicated wretch." The sudden influx of new species in Britain and the drop in prices, however, made them more affordable to collect. Fox agreed to raise the birds, primarily pigeons at first, and Darwin put out a call for pigeon and chicken specimens. That August, as Elizabeth Watts's publication folded, the first red jungle fowl arrived at Down House. Edward Blyth, a magnificently bearded ornithologist, sent "a fine skeleton of the Bengal Jungle-cock" from Calcutta.

Blyth, a brilliant but troubled scientist, had come close to proposing the idea of natural selection himself more than a decade earlier. He publicly defended the views laid out in *Vestiges*, earning Dixon's ire and Darwin's attention and admiration.

After observing all four jungle fowl species during his two decades in Asia, Blyth had definite views on the domestic chicken's ancestor. Despite Linnaeus's classification and studies by Bachman and others, the bird's origin remained an open question that Darwin was eager to answer. Blyth dismissed the Sri Lankan jungle fowl as a candidate. First cataloged in the 1830s by the French naturalist René-Primevère Lesson—who named it *Gallus lafayetii* after the Marquis de Lafayette, the war hero and virtual son to George Washington—its range was restricted to this island off the southeast coast of India. The gray jungle fowl, named *Gallus sonnerati* after the French naturalist Pierre Sonnerat, likewise was found only in a relatively small habitat, in the southern portion of India. Its crow differed significantly from that of the barnyard bird.

Since the green jungle fowl found on the Southeast Asian island of Java lacked the single comb, wattles, and neck feathers typical of a barnyard fowl, Blyth excluded that species as a contender, which left the red jungle fowl as the most likely progenitor. This variety, which includes five subspecies, lives within a vast swath of Asia from today's northern Pakistan to Indonesia's Java, each centered in a different zone of that enormous area. This adaptable bird, Blyth noted, "absolutely & essentially conforms to the type of the domestic fowl in all its multitudinous varieties, fully as much so as the Mallard does to the domestic drake; or the wild to the tame Turkey!" The resemblance extended to the rooster's crow, and the red jungle fowl "corresponds feather by feather with many domestic individuals."

In December 1855, Blyth wrote Darwin from Calcutta with electrifying news about a paper recently published by Alfred Russell Wallace, then living in Borneo. "Wallace has, I think, put the matter well; and according to his theory, the various domestic races of animals have been fairly developed into species," the ornithologist reported. Darwin professed to be unimpressed with Wallace's research, but within a few months set to work on his definitive book on natural selection, quickening his pace of collecting and expanding his contacts. While prowling a poultry exhibition in south London, the naturalist ran into the journalist and poultry fancier William Tegetmeier, and implored him for help in purchasing specimens. Tegetmeier became a key contact and collaborator.

Darwin cast a global net. He importuned a missionary to provide information "in regard to the domesticated poultry and animals of East Africa." In March 1856 he asked his former servant and assistant, who had moved to Australia, to report on "any odd breeds of poultry, or pigeons, or ducks, imported from China, or Indian, or Pacific islands" and send the skins. He dug into chicken history at the British Museum and had a centuries-old Chinese encyclopedia entry on chickens translated to provide clues to the bird's origin. "This morning I have been carefully examining a splendid Cochin Cock sent," he wrote the cousin Fox on March 15 from Down House. "I find several important differences in number of feathers . . . making me suspect quite a distinct species."

His appetite for poultry was enormous. That fall, live chickens arrived from the interior of Sierra Leone, since he feared that coastal birds were not indigenous to Africa. "By the way," he wrote one supplier, "I should be very glad of a Malay Cock, if you can ever get one dead for me." Other live birds and skeletons, including "1 Cock of doubtful origin ticketed Rangoon," began to arrive at the door of Down House. One can only wonder what the postman thought of these strange deliveries. "I expect soon," he mentioned in one letter, "to receive Persian fowls." The shipping, he complained, was costing him a fortune. Post for the Persian birds alone ran to what would now be about five hundred dollars.

By November, Darwin told Tegetmeier that "I really now think I shall have materials to judge the Poultry of the World" in order "to know what amount of difference there is in the Fowls of different parts of [the] world." To try to determine to his own satisfaction the species that was the chicken's progenitor, he began to crossbreed birds to see if Blyth was right in the assertion that the red jungle fowl was the original source. By isolating traits common to domesticated chickens and specific jungle fowl, he would confirm the bird's origin and silence critics like Dixon.

On the Origin of Species, published in 1859, barely mentions chickens, though Darwin gives Blyth credit for the theory that all breeds originated with the red jungle fowl. He put off an extended discussion of artificial selection—that is, domestication—until nearly a decade later. *The Variation of Animals and Plants under Domestication* is less well known than the famous *Origins,* but it lays out Darwin's radical vision of how humans and nature collaborate, sometimes in harmony, sometimes uneasily, and sometimes by accident, to make and remake plants and animals. Humans could not create life from scratch, but they could alter a species by substituting a different climate and soil, giving it food and shelter, and making choices—both conscious and unconscious—about how and when and with whom it would breed.

Darwin tackles the perplexing question of how chickens could have such outrageous variations and still be descended from the same ancestor. Since ancient times, breeders have known how to select the best birds in order to improve their stocks. Fashions came and went in animals as well as clothes. Love of novelty led some Britons to develop "rumpless" fowls lacking a final vertebra, while Indians bred for a frizzled feather look. Some of the novelties would be prized and carefully preserved. Ancient Romans, for example, liked their chickens with an extra toe and white ear lobes that persist to this day. Such breeding didn't require written records or fancy poultry experts. In the Philippines, Darwin notes, "no less than nine sub-varieties of the gamecock are kept and named, so that they must be separately bred."

Human control is limited, and enthusiasts like Dixon had overlooked unconscious or unmethodical selection. Birds escape their coops and mate while farmers are asleep. This takes place even in the best-controlled research facilities. And natural selection, the spinning of the genetic dials, can alter how a creature looks and behaves, a process that continues to function even in the barnyard. The occasional bird will be born with what Darwin calls "abnormal and hereditary peculiarities." Humans might make the most of that new trait, spinning off yet another variety of fowl, or they might decide to not let that characteristic propagate. Darwin summarizes that "there is no insuperable difficulty, to the best of my judgment, in believing that all the breeds have descended from some one parent-source."

The *Gallus gallus bankiva*, the subspecies of red jungle fowl that Wallace collected in Malacca, is the best candidate for the ancient chicken's ancestor, Darwin concludes. The domesticated birds that most nearly resemble this wild fowl, he notes, are the gamecocks of Malaya and Java, which are close in size, color, and skeletal structure and even in their crow to the local red jungle fowl. People in the region keep wild cocks to fight their domesticated ones, and the bird may have been first domesticated in Malaya and exported to India. Chinese texts consulted by the naturalist in the British Library support this idea, since they hint at a Southeastern Asian origin for the domesticated bird. Darwin not only pinpoints the ultimate ancestor of the chicken, he also names the region where humans likely domesticated the wild creature. He ruefully acknowledges, however, that "sufficient materials do not exist" to reveal the complete history of the bird.

Just as Dixon used Darwin's data to argue that species don't change, Darwin used Dixon's data to show the opposite. The rector's experiments demonstrated that crossbreeding other jungle fowl with the red led to infertile progeny. In a later and larger experiment involving five hundred eggs made by crossing gray and red jungle fowl at London's Zoo, the handful that were raised proved sterile, but the red could cross with domesticated chickens and produce fertile birds. "Hence it may be safely ranked as the parent of this, the most typical domesticated breed," Darwin concluded.

In 1868, a few months before *Variations* came out, the naturalist
packed up all his chicken skeletons in his Down House study and
donated them to London's Natural History Museum. "He got rid of
them even before he published—you get the sense he was relieved
to be done with them," Cooper says as she leads me down a flight
of stairs to the skeleton depository, also filled with identical storage
cabinets. She motions for me to wait at a long table against one wall.
Returning, Cooper looks like a waitress bussing a busy cafeteria. She
carries a large tray of a half-dozen chicken carcasses that have been
picked clean.

Each is in its own rectangular plastic case that seems only slightly
more substantial than the plastic cases of whole roasted chickens at
the supermarket. She sets each on the table and goes back for more.
I stare at the bones through the plastic. Some are marked with numbers and names. "That's his writing," Cooper says as she returns to
set down another tray of boxes. Here is the rooster that arrived at
Down House from Sierra Leone a century and a half ago. There is the
Cochin hen, the rumpless fowl, and the Malaysian hen along with
fighting cocks and little bantams. And on the far end is a box with a
skeleton noticeably smaller than that of most of the other domestic
birds. I open the plastic top and peer inside. There's a small note, the
ink still dark and the writing elegant. "Skeleton of Bengal Jungle cock
for C. Darwin Esq.," it reads. It is signed by Blyth. On a leg bone, Darwin has etched in tiny characters a single word, "Wild."

Many poultry fanciers derided Darwin's assertion that a tiny Sebright
bantam and a giant Chinese Silkie descended from a single ancestor. As recently as 2008, one headline insisted "Darwin Was Wrong
About Wild Origin Of The Chicken, New Research Shows." A team
that included Leif Andersson from Uppsala University—his lab is
just down the street from Linnaeus's home—found that yellow skin
in domesticated chickens can be traced to the gray and not the red
jungle fowl. Since that trait might have been added long after initial
domestication took place, Darwin may still have gotten it right.

Pinpointing the chicken's ancestor still left a host of mysteries unsolved. Was it domesticated once or many times? Where? When? And why? A century and a half later, these points remain hotly debated, and reflect larger unanswered questions about humans. For example, if domestication occurred only once and then spread across South Asia and the world, it could tell us that humans were busy trading goods and ideas as well as animals long before the advent of cities and caravans. If, however, chickens and our species forged partnerships in different times and places—Vietnam, Malaysia, India—then prehistoric peoples may have relied more on their own homegrown technologies than those produced elsewhere.

China, India, and Southeast Asian nations all proudly claim to be the original chicken's homeland. Actual ancient chicken bones could settle the dispute. Archaeologists can use radiocarbon techniques to date an old bone to within a few centuries, but the wet and acidic soils of Southeast Asia so far have revealed no chicken remains older than two millennia. The bone and textual data supporting a domesticated chicken at least four millennia ago is found thousands of miles to the west in what is now India and Pakistan.

Comte de Buffon was skeptical that the chicken's origins could ever be settled and Darwin himself was pessimistic. Biologists today, however, are armed with far more powerful tools than his rulers and calipers. Decoding the genomes of living creatures provides long-hidden information on the origin and changes of species over time. The first serious attempt to use genes to unravel chicken history was made in the 1990s by the second in line to the world's oldest hereditary monarchy.

Like his father, the late Japanese emperor Hirohito, Akishinonomiya Fumihito is a biologist. While Hirohito specialized in jellyfish predators, his second son became fascinated at an early age with the chickens that the empress first acquired in the aftermath of World War II to help feed the royal household. After conducting field research in Southeast Asia, the prince and his collaborators extracted sections of red jungle fowl mitochondrial DNA, the genetic code passed down by the female that serves as a historical tracer of a species.

The team's 1994 conclusion was that chickens were domesticated only once, in Thailand. The work formed the basis of Fumihito's doctoral dissertation and an independent research group confirmed these conclusions eight years later, but two decades after that, this theory unraveled. The American ecologist I. Lehr Brisbin noted that the red jungle fowl used as their wild sample came from a Bangkok zoo and likely was a domesticated hybrid.

In 2006, a team led by Yi-Ping Liu of China's Kunming Institute of Zoology found nine separate clades—that is, groups descended from a common ancestor—in the mitochondrial DNA of a large sample of red jungle fowl and domestic chickens. The distribution of these clades suggested several domestications rather than just one. Ancient peoples in southern China, in Southeast Asia, and on the Indian subcontinent separately bred red jungle fowl, they argued, creating distinct lineages with their own genetic signatures. A 2012 study that used nuclear DNA, which provides more detailed data on an organism than the mitochondrial sort, supported the claim of multiple origins of the chicken.

Fumihito himself is convinced by the new data. He could not speak with me, since an interview with any member of the Japanese royal family is difficult to arrange, but a person familiar with his views did say this: "Earlier I thought domestication of the chicken took place in continental Southeast Asia and chickens dispersed from there. Based on recent studies, however, it is more likely that the chicken was domesticated in multiple locations, such as in India, in southern China, and perhaps also in Indonesia. In any case, I don't think it was a single event."

Other geneticists, however, may vindicate the prince's original view. They claim that extensive analysis shows that the chicken indeed originated in Southeast Asia and spread from there to other parts of Asia and then the rest of the world. Olivier Hanotte, a biologist at the University of Nottingham in Britain, and his young colleague Joram Mwacharo have spent the past few years analyzing blood samples from thousands of village chickens across Asia, Africa, and South America. With the help of other colleagues and village children capable of catching the hard-to-get birds, Hanotte and Mwacharo have assembled more than five thousand genetic sequences of modern chickens.

Tracing the chicken's history through the bird's blood is immensely complicated since the bird's genome is scrambled from being moved back and forth across oceans and continents for millennia, its sequences mixed and remixed. In their Nottingham lab, Hanotte and Mwacharo show me a PowerPoint presentation that lays out that complexity. Their genomic maps show a half-dozen major haplotypes, or genetic groups, connected by a confusing array of arrows and bars. "How do you reconcile huge diversity with a single domestication?" asks Hanotte, a rangy man who speaks rapidly with a musical accent. The complexity of the chicken genome points to multiple domestications in different places.

Hanotte believes that this complexity actually stems from several wild subspecies of the red jungle fowl mating and occasionally producing hybrids. Three of the five subspecies of red jungle fowl are Southeast Asian. Also, the greatest variety of domesticated fowl are found there as well. As one moves west, toward the red jungle fowl's western terminus in Pakistan, those distinct varieties decrease. This is why chickens from Britain's lands on the Indian subcontinent didn't dazzle Queen Victoria, fascinate Dixon, or intrigue Darwin the way that those from Southeast Asia and southern China did.

This genetic diversity in both wild and domesticated varieties is a strong indicator that this region is the original homeland of chicken domestication. Hanotte and Mwacharo also pinpointed what biologists call a "population bottleneck" in the same region. That is, the bird's genetic diversity suddenly decreased in the distant past, a sign that an early domesticated bird radiated from some point in Southeast Asia across the region. Hanotte estimates that the bottleneck began from eighteen thousand to eight thousand years ago. The newly domesticated bird may have moved in small numbers to villages in Southeast Asia and eventually spread to the Indian subcontinent. Then, about five thousand years ago, he believes, the chicken population increased rapidly throughout South Asia and then beyond.

This suggests that the bird and humans originally made common cause for very different reasons than why we keep chickens today. At first, the domesticated animal was not bred for meat and eggs. "There does not seem any ground for believing that an attempt was made at this early period to domesticate them for purposes of food," a group of poultry scholars noted as early as 1854.

The indigenous peoples of Southeast Asia retain a remarkably rich variety of traditions concerning the fowl. The Palaung people scattered through Vietnam, Laos, Burma, and southern China, for example, keep chickens but don't eat the meat or eggs, instead using the bird's intestines, organs, or bones for divination. Likewise, the Karen of northern Burma believe that they contact the powerful realm of invisible spirits by the angles that bamboo splinters make when inserted into a slaughtered chicken's thighbones. The Purum Kukis, who live in India's far northeast, close to Burma, recite prayers, sacrifice a cock, and judge by the way it falls whether the spot is a good place to build a village. Many tribes across the area practice an egg toss in which the resulting shell fragments will provide clear guidance for thorny situations. These beliefs in the magical nature of chickens may predate the arrival of agriculture, when people began to use the bird more commonly for food.

At Uppsala University, Andersson and a team of collaborators recently found that several different populations of chickens share a mutant form of a gene absent in the red jungle fowl. This bit of DNA produces a particular hormone that stimulates the thyroid. A bird with this code could gain weight faster. This mutation may have been an important step in the evolution of the chicken. In some South Asian village long ago, a bird born with a proclivity to grow faster—and to procreate more frequently—was singled out for special breeding.

But chicken as food may have been an afterthought. Long ago, when we first took or enticed it from its forest home, the bird was more than a cheap lunch. It was magical and practical at the same time. Along with its powers of divination, its delicate bones could be used for sewing or tattooing or to make small musical instruments. The rooster's magnificent feathers could adorn ornamental clothing.

The bird had well-known medicinal properties and its martial abilities provided entertainment. As a sacrifice to the gods, the small and fast-reproducing animal was ideal.

Amid familiar mammals like dogs and cats and cows, however, the chicken retains an almost alien quality. The male can be fierce and even terrifying when defending its turf, all out of proportion with its small size, while the bird's reptilian feet and downy feathers make for an unsettling combination. Jerky movements give the bird a disturbingly robotic quality, while its voracious appetite for sex with multiple partners is impressive to some and offensive to others. We veer between admiration and disgust and between fascination and fear in our long relationship with the chicken. This ambivalence mirrors our shifting attitudes toward God, sex, gender, and all that we consider both sensual and monstrous.

8.

The Little King

Why are the cocks not crowing, he muttered to himself, and
repeated the question anxiously, as if the cocks' crowing
might be the last hope of salvation.

—José Saramago, *The Gospel According to Jesus Christ*

The cock has no cock. That's no Zen Buddhist riddle. The rooster
lacks a penis. Or, to be more accurate, it lost it, which is a rather odd
fact for the animal that more than any other is the zoological stand-in
for the human male organ. The mystery of how the chicken phallus
vanished was recently solved, although why it vanished remains a
matter of some controversy among the small but dedicated band of
bird-penis researchers.

When chickens mate, a casual spectator might be forgiven for
thinking that fowl sex is similar to that of mammals. The rooster
mounts the hen, takes a firm grip of her back with his claws, and uses
his beak to hold on to her head. Then he hops off. It is usually over
in less time than a Filipino cockfight. Though it can be a similarly
raucous affair, two birds have exchanged a cloacal kiss.

The cloaca, from the Latin word for sewer, is a busy place. In the
chicken—as well as all birds, reptiles, and amphibians—the cloaca

is the single-lane end of the urinary and digestive tracts, and it also serves reproductive duties. Like human males, roosters have two testes, though these are internal organs tucked under the kidneys rather than dangling on the outside. A healthy rooster can produce more than 8 billion sperm in every ejaculation that are transferred into the hen when they invert their cloacae and press them together. It only takes a few seconds. The sperm in her oviduct can fertilize eggs in her single ovary for up to a month after they mate.

A few species of birds, mainly waterfowl, do have penises. For example, ducks have long corkscrew-shaped peckers. A team overseen by Martin Cohn at the University of Florida, Gainesville, investigated why ducks and chickens differ in this respect. They cut small windows into eggs to view both male duck and chicken embryos, and found that both began to develop penises for the first nine days of development, after which growth in chickens stopped and the proto-organ shriveled up. On day nine, the chicken embryos began to manufacture a protein responsible for shriveling the would-be penis from the tip down. That protein also is tied to the loss of nascent chicken teeth during early development and it influences beak shape and feathers. It is essentially a chemical that kills off selected cells. When the researchers coated the would-be chicken penis with another protein that blocked the shriveling effects, it kept growing. Coating a duck penis with the cell-killing chicken protein reversed its growth.

Cohn thinks this organ loss was simply a consequence of dropping other body parts like teeth and limbs. This particular disruptive protein clearly plays a major role in bird evolution, and losing the penis was just a side effect. Other biologists suspect that the change is the result of female selection, which led to the evolution of the cooperative cloacal kiss over rougher penetrative sex. Male ducks are notorious for forcing copulation on uncooperative mates and sometimes even drown their mates in the process. Such battles of the sexes reduce the odds for fertilization. Over time, most bird species, including the chicken, may have selected males with smaller penises while others, notably ducks, geese, and swans, kept theirs.

How the penis vanished in chickens—and most birds—could tell us much more about evolution, such as how snakes lost their legs and what causes birth defects in human genitalia, which are particularly susceptible to malformations in the womb. The research conceivably could lead to practical ways to correct those defects before birth. "Genitalia, dear readers, are where the rubber meets the road, evolutionarily," wrote the biologist Patricia Brennan from the University of Massachusetts, Amherst, in *Slate*. She was responding to an outcry by some media commentators outraged that federal tax money would be spent on chicken penis research. "To fully understand why some individuals are more successful than others during reproduction, there may be no better place to look."

Losing the penis also might ultimately have given the chicken a slightly higher fertility rate than those other common farm birds outside the family of Galliformes. But in losing its cock, it is possible that the chicken may have gained the world.

Biology can't explain why our favored slang word for the male organ refers to a bird that lacks one. Americans blush at a word bandied about shamelessly by Canadians, Australians, British, and other English speakers, who still use it without hesitation when describing the male chicken. Eighteenth-century New England Puritans excised *cock*, a word that likely derives from the sound of a chicken—it comes from the ancient Aryan word *kak*, which translates as "to cackle"—from the American lexicon.

These were, after all, people who punished those wicked souls who celebrated Christmas. Puritans did not oppose sex and were harsh critics of celibacy for Roman Catholic clergy, but double entendres led the mind and body astray and therefore were not to be tolerated. Two centuries before, in Elizabethan times, one poem begins with the line "I have a gentle cock" and ends with: "And every night he perches in my lady's chamber." This bawdy tradition continued for centuries. The 1785 *Classical Dictionary of the Vulgar Tongue* notes that the phrase "cock alley" means "the private parts of a woman."

In the American colonies, the more anodyne *rooster* began to replace *cock* just before the revolution, starting in the north and spreading south. The term derives from an Old English word for the perch of a domestic fowl. In the young United States, *haycocks* became *haystacks*, *weathercocks* became *weathervanes*, and *water cocks* became *faucets*. Even *cockroaches* turned into just plain *roaches*. "Victoria was not crowned in England until 1838, but a Victorian movement against naughty words had been in full blast in this country since the beginning of the century," H. L. Mencken notes sardonically in *American Language. Cock* had to go because it had acquired "an indelicate anatomical significance." In Britain, the word continues to be used in all its glorious ambiguity. Well into the Victorian era, *cock* was still preferred among British doctors as a descriptor of the human organ over the more newfangled term *penis*, adopted from the French but originating in Latin.

The cock most likely acquired this "indelicate anatomical significance" due to extremely randy behavior—and research demonstrates that the male chicken prefers new partners to familiar ones. Scientists call this salacious behavior the "Coolidge effect." During separate tours of a chicken farm by President Calvin Coolidge and his wife in the 1920s, Mrs. Coolidge remarked on a rooster that was busy mating. She was told that this behavior took place dozens of times daily. "Tell that to the president when he comes by," she said coolly. When the message was relayed, the president asked if the rooster mated with the same hen. He was told no, that the male preferred a variety of partners. "Tell that to Mrs. Coolidge," he responded.

The chicken's reproductive vigor has long impressed humans. The Babylonian Talmud, compiled in the early centuries AD in today's Iraq, mentions the custom of Jews carrying a hen and cock before a bridal couple, a tradition that continues in parts of the Middle East. This ritual was not just about fertility. The Greek god Zeus gave the handsome Ganymede a live rooster, and older aristocratic Athenians in Aristotle's day presented their young male lovers with such a bird. There is no mistaking the sexual reference at Delos, near an ancient temple to Apollo, where a large column dating to Aristotle's day supports an anatomically correct massive erect human penis and testicles. Just below is a carved rooster, its head and neck in the form of a phallus. "Throughout classical antiquity," writes the art historian Lorrayne Baird, "the cock serves as icon and symbol of the male erotic urge."

In classical art, roosters pull Eros's chariot, watch Mars and Venus make love, or observe Menelaus seizing Helen of Troy. In a Berlin museum, a Greek vase dating to 500 BC is decorated with a line of black-robed men dressed as roosters in a drama chorus, following a piper. Nearby stands a small bronze statue of a rooster man, dating to early Roman times and dug up near Mount Vesuvius. He has a high swept-back comb that gives him a punk-Mohawk look. The lips are drawn back in pleasure as enormous wattles droop down to his breast. He has pushed aside the fabric around his waist to expose a massive phallus as long as his torso, held aloft by his right hand and the sheer force of the erection.

Hidden from public view in the Vatican archives is a small bronze bust of uncertain date. It has the upper torso of a human male and the head of a rooster. In place of the beak is a massive and anatomically exact penis that takes up most of the face. Inscribed in Greek on the base are the words SAVIOR OF THE WORLD. For a century or so it was on display until an appalled eighteenth-century cardinal complained. Its authenticity has been questioned, but a similar bust now in a German collection was found in an ancient Greek temple, suggesting that this rooster-as-savior concept dates back at least to Socrates's time.

Lots of farmyard animals—goats and dogs come to mind—are frisky, but along with the rooster's sexual prowess comes its unique announcement of the coming of the light. To the ancients, dawn was a religious event tied to the creation and re-creation of life itself. "In Nature man generates man," Aristotle wrote, "but the process presupposes . . . solar heat." Associated with the rooster is a long list of solar gods, from the Greek Apollo and the goddesses Leto and Asteria to the popular Roman-Persian deity Mithras and Zoroaster's Ahura Mazda. The piece of pottery from Egypt, the first clear depiction of the bird, that Carter found in the Valley of the Kings, might be tied to the solar cult of Akhenaten, while the image on the ivory box in

Mesopotamia's Assur hints at the bird's connection with the Shamash deity later favored by Babylon's last king, Nabonidus.

By the time of Christ, roosters were common images in temples from Persia to Egypt to Britain and their sacrificed remains show up in tombs from Turkey to Britain. "They declare the cock to be sacred to the sun, and the herald to announce the coming of the sun," declared the second-century AD geographer Pausanias while passing through southern Greece. Even among Jews, who long considered chickens unclean animals, the rooster crow was the signal to say a benediction. "Praised be to the Lord my God for giving the cock the intelligence to distinguish between day and night," goes an old Jewish morning prayer. In ancient China and Japan, the cock symbolized the sun. "Lo, the raving lions, / They dare not face and gaze upon the cock / Who's wont with wings to flap away the night," wrote Titus Lucretius Carus in his first-century BC treatise *On the Nature of Things*.

Given this wealth of tradition, the rooster's early emergence as the single most important animal symbol of Christianity makes more sense. Jesus is associated with the lamb and the fish, the Holy Spirit with the dove. The lion, ox, and eagle represent three of the four gospel writers. Even peacocks were borrowed to symbolize saints. But over all of these, often even above the highest cross, atop thousands of steeples and domes around the world, a weathercock glints. These are ostensibly reminders of Jesus's warning to Peter that he would deny his teacher three times before the cock crowed twice on the day of Christ's crucifixion. But the bird's heralding of light and promise of resurrection are deeply interwoven in Christian tradition. Jesus's birth, Peter's denial, and the Easter resurrection were said to have taken place at the cock's crow. In a popular fourth-century AD hymn, the bird is called "the messenger of dawn" that, like Jesus, "calls us back to life" and wakes those who are sick, sleepy, and lazy. Like Christ, the rooster restores health, awareness, and faith. "God, the Creator Himself, becomes a sort of divine cock," writes Baird.

The connection between the rooster and the new religion was strongest at the heart of Western Christianity, in Rome, where tombs of early converts were carved with fowl engaged in sacred combat. Peter was crucified on Vatican Hill in the first century AD, and one historian suggests that he assumed the doorkeeper role of the old Roman god Janus. According to ancient Etruscan beliefs, Janus was the sun and the cock was his sacred bird of dawn. The key symbolizes this deity, whose name means *archway* in Latin, and who stood guard at the gate of heaven. "I will give you the keys of the kingdom of heaven," Christ told Peter, whose name means *rock*, in the gospels.

The bird's link with the Vatican may even have more esoteric roots. Today's Saint Peter's Basilica sits on the site of a temple to Cybele, the great mother goddess of Anatolia, who became popular in Rome as Christianity began to take root. Unlike Janus or Peter, she held the keys to the underworld rather than heaven. Her priests were castrated men called Galli, a reference either to the Roman word for *rooster*, an ancient king, a river in Anatolia, or some combination of those. "The erotic association of 'rooster-cock-phallus,' already in currency, served as an inside joke among the Galli and was later used by Roman citizens to ridicule them," one scholar notes.

The poet Juvenal remarked that Galli were "capon'd late," that is, had their testicles removed as was done with some roosters. The gender-bending Galli dressed and behaved like women, and the shrine remained venerated well into the fourth century AD, possibly after construction of old Saint Peter's Basilica by the Roman emperor Constantine I. Near both the original basilica and the Cybele temple stood the great obelisk brought to Rome from Egypt's Heliopolis and rededicated to the sun god by the Roman emperor Caligula in 37 AD, three decades before Peter was crucified in its shadow.

By the late sixth century, Pope Gregory I decreed that the rooster's tie to St. Peter made it Christianity's most suitable emblem. (The name Gregory comes from the Greek word for *vigilant* and may have been related to the bird.) "The Cock is like the Souls of the Just, waiting for the dawn, after the darkness of the world's night," wrote the English monk and scholar Bede a few decades later. By the ninth century, a huge gilded rooster shone from the bell tower of old Saint Peter's Basilica, calling for the faithful to awake. Clergy were called

"Cocks of the Almighty." By the tenth century, by papal decree, every church in Christendom was ordered to place a rooster on its highest point.

In 1102, at the end of the First Crusade, Europeans in Jerusalem rebuilt a destroyed fifth-century AD Byzantine shrine where tradition says Peter's betrayal took place—the house of the priest Caiaphas—and named it the Gallicantu, or the cock's crow. Inside the building, reconstructed in the 1930s, is a painting of Jesus and Peter standing on either side of a slender column surmounted by a rooster. In the courtyard outside is a statue commemorating the gospel scene, with a rooster standing high above on a stone column that eerily resembles the images on ancient Babylonian seals and Greek vases.

By the time of the First Crusade, the rooster was falling out of favor among the clergy, which increasingly saw it as licentious, subject to the temptation of lust. Yet the bird remained popular with common people as a protector from evil. Amulets and incantations invoking the bird remained in use despite church bans on them. Magical amulets displaying fierce creatures with snake legs and a cock's head date back to Greco-Roman times, and were popular among ancient Jews and Persians as well as medieval Christians. The rooster's long history as a power animal could not be quickly erased.

In the fourteenth century, swearing often substituted *God* or *Christ* for *cock*—Chaucer's characters spoke of "cock's bones." Alchemists, those protochemists, sought a powerful stone said to reside inside a rooster's head that "will cause you to obtain whatever thing"—from an eloquent tongue to a woman's ability to please her husband sexually. They often used the rooster as a symbol of the sun and the masculine principle, and the hen for the moon and the feminine. Observing chickens, they began to assert that the female was more than just fertile ground for the life force of the male seed, an idea considered doctrine since Aristotle. This radical notion, which went against church teaching, helped lay the foundation for the modern science of embryology.

A last echo of the cock's Christian role is heard in Shakespeare's *Hamlet*, written around 1600. In the Christmas season, "the bird of dawning singeth all night long," says a sentry in the Danish castle of Elsinore. "And then, they say, no spirit dare stir abroad; the nights are wholesome." A decade later in the town of Pagani, south of Rome, a church was dedicated to the Madonna of the Chickens after several hens pecked on a wooden icon of the Virgin and miraculous cures ensued. But the Reformation knocked the teetering rooster off its pedestal. By the seventeenth century, witches were accused of using the bird as a substitute for the host at illicit black masses. In a dramatic reversal, the animal was seen as a harbinger of darkness and a tool of Satan, often pictured in league with the devil or as the devil itself. Protestant artists working on murals in the inner court of the city hall in Basel, Switzerland, painted a vivid scene of hell, with Satan as a human-sized rooster torturing evildoers, including a pope and a nun.

"Within the Cock, vile incest doth appeare," warned the English curate Henry Peacham in 1612, as well as cautioning against sodomy, witchcraft, and murder. The cock's comb that once inspired the serrated crown of ancient Persian kings was instead remade as a jester's silly hat that went by the same name. In *The Taming of the Shrew*, written after 1590, Shakespeare uses this change for a spirited and ribald exchange. "What is your crest?" Kate demands of her would-be wooer Petruchio, asking facetiously about his family's heraldic display. "A coxcomb?" But Petruchio is quick with a snappy comeback: "A combless cock, so Kate will be my hen." A comb could also be interpreted as the foreskin of an uncircumcised penis. Some bard scholars believe Petruchio is implying his foreskin isn't showing because he has an erection in the presence of his prospective bride.

At the dawn of the modern era in the West, the once proud and sacred bird remained linked to sex, but increasingly as an emblem of lust and ridicule. In the seventeenth century, a deep strain of sexual puritanism—both Protestant and Catholic—spread across Europe and into the Americas. When old Saint Peter's was demolished— exposing remants of the Cybele shrine—the bronze cock that for a millennium preened from its bell tower was relegated to the Vatican treasury, where it remains today. The new basilica was consecrated in 1626, with Caligula's obelisk placed at the center of Saint Peter's Square. Thirty years later, citing "superstitions among the people," Pope Alexander VII banned an ancient ritual in which a new pope received a bronze cock that sat atop a porphyry column at Rome's Saint John Lateran, said to be the very one from which the cock crowed Peter's denial in Jerusalem.

By this time, clock makers were at work perfecting mechanical alarm clocks. Soon the rooster's call would be more nuisance than blessing. Despised for its ungodly ways on both sides of the religious divide in the West, the cockless cock was now neither royal nor sacred. Even so, the male chicken could not be completely eradicated from public life, even in New England. The Puritan preacher Cotton Mather dedicated a Boston church in 1721 that came to be known as "the Church of the Holy Rooster." The five-foot-five-inch, 172-pound gilded weathercock became the city's preeminent landmark for sailors. By the time it blew down during a storm in the late nineteenth century, it had helped guide clipper ships from China carrying exotic fowl into Boston harbor.

↞ ↞

Chicken is synonymous with cowardice, a twentieth-century idea that would bewilder the ancients, puzzle our more recent ancestors, and annoy the French. The rooster remains, after all, the national emblem of France. It also has a longer history than the donkey as a U.S. Democratic Party mascot. Groups as diverse as the Communist Party of Venezuela (which uses a design drawn by Pablo Picasso), Robert Mugabe's Zimbabwe African National Union, and Berlin's Protestant student union use it as their symbol. In our modern world of factory-farmed birds, however, the cock's martial abilities are no longer needed, appreciated, or desired. Yet the bird's natural fierceness has formidable roots.

In 2007, a scientific team extracted a protein from a dinosaur that lived 68 million years ago and found it to be identical to one that exists in the chicken. This was not just any dinosaur but the largest two-legged carnivore of all. "Study: *Tyrannosaurus Rex* Basically a

Big Chicken," read one headline. Paleontologists in the past decade have accepted the idea that birds evolved from dinosaurs, but the protein discovery marked the first time that biologists had genetic proof of the connection.

The discovery began in the rugged badlands of northeastern Montana. Jack Horner, a largely self-taught Montana paleontologist, was leading a team sampling the rich fossil fields in the area. Under tons of debris and rock they recovered a remarkably intact *T. rex*, including a three-foot-long femur. The fossilized remains were encased in protective plaster, and the one-ton weight proved too heavy to lift by helicopter, so the team sawed it in half. In the process, the femur broke and shed a few pieces. In 2003, Horner shipped these bits to his former student Mary Schweitzer at North Carolina State University in Raleigh, who used molecular biology to analyze dinosaur remains. Unlike bone, tissue quickly degrades, so she didn't anticipate finding any in the sample.

Schweitzer noted the femur belonged to a pregnant female, since there was a particular kind of tissue within the bone made during ovulation to conserve calcium. It was the first time anyone had found unimpeachable evidence of a dinosaur's sex. The following year, she asked one of her assistants to soak a fragment in a weak acid. Since fossils typically are mostly rock that dissolves quickly in such a solution, the procedure can destroy a sample, but the assistant discovered a rubbery substance left behind after a long soaking. When other fragments were subjected to identical treatment, the same material was left behind. The two researchers could even make out what looked like blood vessels. Schweitzer had found the first dinosaur tissue. Unlike *Jurassic Park*'s scenario of cloning dinosaurs based on blood found in a mosquito trapped in amber, there was no chance of recovering DNA from the samples, but the tissue held other secrets.

A Harvard University chemist named John Asara had worked with Schweitzer a few years before, identifying the proteins from a three-hundred-thousand-year-old mammoth bone, although he specialized in sequencing proteins in human tumors. Proteins are made up of amino-acid chains too tiny to be imaged by a typical lab microscope, but Asara knew how to add antibodies that would bind to the proteins and make them visible.

Schweitzer sent Asara via FedEx a little vial of brown powder wrapped in dry ice, the ground-up soft tissue from the femur fragment. He carefully removed the brownish contaminants in the powder. "You don't want to inject anything brown into a three-hundred-thousand-dollar machine," he explains when I visit him in his Harvard lab in a Boston high-rise. The mass spectrometer is a boxy plastic device the size of a hotel-room refrigerator that can measure the tiny masses and concentrations of molecules and atoms.

Asara first added an enzyme to break up any proteins into more manageable molecules called peptides. The mass spectrometer then spit out nearly fifty thousand spectra detailing the composition of the sample. There is no database on dinosaur DNA sequences, so Asara had to make theoretical models of protein sequences that *might* have existed 68 million years ago, based on earlier mastodon work. He also had the sequence of the chicken that had been published in 2004. "We have a more comprehensive database for chickens than any other bird," he says.

He pinpointed a half-dozen protein lines in the *T. rex* that were identical matches to the chicken. Not only had he and Schweitzer isolated 68-million-year-old soft tissue, more than twenty times older than any yet found, they also asserted that they had identified the world's oldest proteins and found them identical to those in the modern chicken. Their 2007 paper in *Science* clinched the argument for placing birds with dinosaurs in the evolutionary tree, though skeptical colleagues tried to rebut the claim. Two years later, Schweitzer and Asara found eight sequences of chickenlike proteins in an 80-million-year-old bone of a hadrosaur, vindicating their technique.

Reverse evolution can provide more understanding of the link between dinosaurs and modern birds like the chicken. Horner, the Montana paleontologist, proposes creating what he calls a chickenosaurus by peeling back the genetic layers of the chicken to expose the monster within. The vanishing embryonic chicken penis is an example of how evolution is revealed in embryo development. Fetal chickens also briefly begin to grow a dinosaur-like three-fingered claw and a long tail that then disappears. In theory, if molecular biologists could prevent the gene that gets rid of the tail from turning on, then a hybrid chicken-dinosaur could result. Genes from other species could also be added to encourage dinosaur-like traits and repress chicken ones.

Mutant chickens might also provide insight into bird-dinosaur evolution. In 2004, a biologist working with a chicken embryo found tiny bumps inside the developing mouth. These were not the flat-topped enamel that dominates human jaws, but sharp and conical structures resembling miniature alligator teeth. The researcher went on to create a virus that could copy the signals sent by the genetic mutation and set off similar teeth growth in normal chicken embryos. The teeth didn't last and were absorbed into the beak, but the experiment offered a glimpse into that long-lost era when hen's teeth weren't rare.

Another recent experiment produced crocodile-like snouts on chicken embryos. Arkhat Abzhanov, an evolutionary biologist with glasses and a trim black goatee, conducts the research at his Harvard lab. "I'm not sure that the goal of re-creating a 'dinosaur' is necessarily a good science project," he says when I stop by his office. Abzhanov runs what may be the world's only chicken boot camp, in which young researchers spend six grueling weeks learning how to make the best use of eggs for the greater glory of biology. "They are truly a fantastic system."

Chicken embryos are tough and large and very predictable. They can be stored in a cooler for two weeks without developing. In an incubator, their changes from hour to hour have been well mapped. What begins as a little disk that is a mere two cells thick turns into an intricately structured and complex organism. Cut a hole in the shell, cover it with clear duct tape, and you can watch it happen. Eggs are also cheap and easy to store and manipulate, unlike primates, mice, and even zebra fish. So long as he doesn't hatch them, Abzhanov is free to do what he wants with chick embryos since they are outside lab-animal regulations. His entire lab is a small, windowless room with a long counter, a couple of microscopes, empty egg cartons, and a clock featuring a rooster. A grad student is at work injecting a virus expressing a certain protein into embryonic tissue; its purplish-brown stain will spread as the embryo cells grow and multiply, providing a trail for her to follow in the coming days. This delicate operation, she says, is the toughest part of a boot camp.

Abzhanov grew up in a big city in Kazakhstan in what was then the former Soviet Union, and his earliest memory is chasing chickens at a cousin's farm. As a biologist, he grew fascinated by the evolution of bird heads. He extensively studied Darwin's finches from the Galàpagos Islands to understand what genes were central to the development of their remarkably varied beaks, which make it possible for them to rule a particularly ecological niche. That led him to wonder how the beak developed from dinosaur snouts in the first place, so he turned to chickens and alligators. In the reptile, there are two bones that structure the snout, but in chickens the two are fused. Abzhanov set out to find and turn off the gene that orders a beak instead of a snout to form on the fifth day of a chick's gestation. In 2011 he succeeded—though they were not allowed to hatch because of ethical guidelines.

Short of an ancient DNA sample, you can't "go back" to dinosaurs using chickens. Dinosaurs had many skull features that would have developed later in their embryonic cycle, and that information has been erased in chicken eggs, which develop much faster than, say, those of a *T. rex*. Abzhanov is curious about how dinosaur genes differ from those of a chicken, but his goal is both grander and more practical. "I want to find a more mechanistic view of evolution and disease," he says. Studying the way bird heads develop, for example, might lead to a way to switch off the gene that creates a cleft palate in a human embryo.

Increasingly, dinosaurs are starting to look birdlike without any genetic manipulation at all. In 2007, paleontologists spotted quill knobs on a velociraptor. Four years later, chunks of 75-million-year-old amber revealed actual preserved dinosaur feathers with hints of pigment. These may have been more for display than flight. Triceratops, that hulking, horned, four-legged dinosaur, had feathers on its tail. Even *T. rex* may have been feathered. Feathers, in fact, may have come before flight. Why did these massive animals go extinct 66 million years ago, while the chicken's ancestor survived and thrived?

All birds now are classed among the theropod dinosaurs, which appeared more than 200 million years ago as the planet's first large meat eaters. Even in that early period, some had feathers, hollow bones, and a wishbone still retained by modern birds. Avians are a particular sort of theropod called maniraptorans, a group of long-armed, three-fingered creatures that include velociraptor and *Microraptor gui*, which lived in trees. Another maniraptoran is Oviraptorosauria, which had a beak-like toothless jaw and feathers and produced only one egg from an oviduct at a time—like birds and unlike reptiles—and then sat on its nest to incubate its progeny. Some researchers even class oviraptors, which weighed less than a hundred pounds, as birds.

Abzhanov believes that birds survived and thrived better than their larger cousins because of their small size, which made them more adaptable and less vulnerable. Today's chicken lacks teeth and the huge talons of its massive ancestors, but deep within its DNA, it retains a reptilian ferociousness that—like our fascination with dinosaurs—both appeals and frightens.

For a thousand years in Europe, people believed that the chicken could on rare occasions turn deviant and deadly. A cock could lay an egg that hatched a fearsome monster. Called a basilisk, the rooster-headed creature with the body of a snake or dragon could kill with a glance. Recently, biologists discovered that a cock laying an egg is more than a superstitious myth.

Basilisk means "little king" in Greek. In Roman times, Pliny the Elder called it "a snake with a light crown upon its head," which may describe a hooded cobra imported from India in the first century AD. As the rooster gained ascendancy in Christian symbolism, the basilisk morphed into a more satanic beast combining aspects of the cock and snake long associated with healing in the classical world.

The basilisk haunted the Middle Ages. The twelfth-century German mystic and naturalist Hildegard von Bingen warned that "nothing living is able to endure it" since the basilisk is ruled by the Antichrist. Panic swept thirteenth-century Vienna when word passed that a basilisk was loose in the city's twisting streets. A hysterical mob of sixteenth-century Dutch villagers strangled a rooster and crushed the eggs it was incubating. Warsaw's senate met in an emergency session when several people died gruesomely in a cellar during a basilisk attack. It was destroyed when a condemned prisoner donned a suit of mirrors to defeat the creature. The only thing said to frighten the monster, besides its own reflection and a weasel, was the crowing of the cock.

Fear of the basilisk led to one of history's most peculiar trials. On an August afternoon in 1474 in the Swiss city of Basel, a judge declared an eleven-year-old rooster guilty of laying an egg and ordered it beheaded and burned at the stake. After the executioner chopped off the cock's head and opened its innards, the officials were horrified to find three additional eggs waiting to be laid. All four, along with the corpse, were laid on the fire. Ironically, the basilisk was the city's emblem. The monster, even today, is everywhere in this ancient city. Basilisks spurt water in fountains, open their wings at the entrance to the Wettstein Bridge, and perch on the helmet of a gilded Renaissance statue. The popular local beer is even called basilisk.

As late as 1651, members of the royal Danish court in Copenhagen panicked when a castle servant reported a rooster laying an egg. King Frederick III, an amateur naturalist, kept a cool head and had the egg closely watched. When it did not hatch, he had it placed in his collection of curiosities rather than destroyed. A century later, the basilisk was nothing more than a silly fairy tale. "Know, Sir, that there is no such Animal in Nature as a Basilisk," the sensible character in Voltaire's 1747 book *Zadig* tells the Queen of Babylon. Her retinue seeks the beast to cure the ailing king's health. The basilisk has resurfaced more recently without its rooster looks, in *Harry Potter and the Sorcerer's Stone*, as pure reptile.

At the Roslin Institute in Scotland, which has been at the forefront of finding ways to manufacture protein-based drugs in chickens, the biologist Mike Clinton is the resident chicken expert. A beefy man with a heavy Scottish brogue, he grew up on a remote Hebrides island, where he gathered eggs from his grandmother's farm and cut peat for fuel. Clinton wanted to be a veterinarian, but preferred studying animals to euthanizing them. Fascinated with what determines sex, he used embryonic chicks to see what turns them into roosters or hens.

In 2001, a poultry inspector phoned the institute about a very strangely built fowl that he had just acquired in the south of England. The man had encountered a farmer's son playing with a pet chicken named Sam. Whether that moniker was short for Samantha or Samuel depended on which side of the bird you addressed. Sam's left side had the hulking body of a rooster with white feathers, complete with large comb, wattles, and spur. On the right side, Sam was pure dark-hued hen. Intrigued, Clinton agreed to take the bird, thought to be a one-in-a-million specimen. Two weeks later, the same inspector called again with news that he had found two more.

Sam and the other birds were bilateral gynandromorphs, an animal containing distinctly male and female parts on separate sides of the organism. Lobsters and fruit flies and butterflies occasionally turn up with aspects of both sexes, but this is very rare in vertebrates. Unlike hermaphrodites, which can have sexual organs from both genders but otherwise appear to be either male or female, gynandromorphs are true sexual mosaics.

Scientists have been struggling since Aristotle to understand the mechanisms that determine whether an animal is male or female. The Greek philosopher believed the hotter the sex a man had with a woman, then the greater likelihood that a resulting fetus would be male. This is not as absurd as it sounds, since temperature can play a role in sex differentiation among some animals. The hotter the nest during incubation of alligator eggs, for example, the more likely that males will result.

By the twentieth century, scientists determined that the key to most animals' sex was in the sex chromosomes that take up a chunk of our genetic hard drives. The overwhelming majority of men have one X and one Y, for example, while females have two Xs. One part of the Y chromosome induces the human embryo to create testes rather than ovaries. Those organs can then manufacture and secrete the chemical signals called hormones—testosterone from the testes and estrogen from the ovaries—that tell other cells to produce male and female traits. Before the gonads decide whether to be testes or ovaries, human cells could become either sex.

Many reptiles and birds—and at least one mammal, the platypus—operate slightly differently. Females have a Z and a W chromosome, while males have two of the same kind—Z and Z. Clinton and other researchers had been trying without success for decades to find the equivalent of the human sex-determination gene in birds, the mechanism that makes them male or female. That's why the three gynandromorphs were a surprise and welcome gift.

For two years, Clinton and his colleagues kept the three unusual chickens in a shed isolated from the other Roslin birds. Sam behaved like a male, while the second gynandromorph acted like a female. Two of them appeared female on their right and male on the left, while the third was a mirror image. The third also did not seem to favor either traditional rooster or hen behavior. Sam had testes on his male side, while the second bird had an ovary there instead. The third had a testes-like organ on its female side. After the researchers tried without success to culture living cells from the birds, a veterinarian put them to sleep by injection. In the autopsies that followed, the team took hundreds of tissue and blood samples from both sides of each bird. When Clinton chopped up Sam's testes, he found that the bird made healthy sperm, although his plumbing wasn't up to delivering it into a female and its ovary did not produce eggs. Other gynandromorphs, Clinton says, could look like a rooster and lay eggs.

Thanks to a new dye technique developed as part of Roslin's work in protein-drug creation, the team could color code the tissue according to each Z and W cell. Clinton expected to find that one side of each bird would be normal, while the other side would show some sign of chromosome damage or mutation. To his surprise, each bird had predominantly male cells on one side—Z and Z—and female on the other—Z and W—while they shared blood mixing both.

He realized that almost every cell in these chickens had their own sexual identity, depending on which side they fell. Since the hormones coursing through the birds' bloodstream were the same, it was clear that the cells on each side were marching to a different drummer unrelated to testosterone or estrogen, long assumed to be the real drivers in sex differentiation. It was a chicken-or-egg question. If hormones drive sex determination, then how could they do so before the organs that make them even exist? To understand what was going on, Clinton and his colleagues conducted hundreds of experiments on chicken embryos. The team implanted female chicken cells in male host cells, and males in females. The stubborn implanted cells refused to change their sex role. That means that the fowl is more set in its sex than humans, who go through a unisex phase.

Sam was rare, but not a mutant. Instead the bird was a consequence of an abnormal ovum with two nuclei fertilized by two sperm that develop into two distinct halves of the opposite sex. Clinton's surprising conclusion, published in *Nature* in 2010, was that chicken cells keep their own sexual identity no matter the hormone the animal produces. A chicken's sex is fixed even before testes and ovaries produce testosterone or estrogen to make, for example, a comb and wattles. Unlike with humans, the sexual identity of the chicken is imposed at the moment of fertilization. The Roslin gynandromorphs may shed light on our own species, however. "Even with humans," he tells me, "I think that male and female differences may be independent of hormones, and are the result of inherent differences in male and female cells." This might prove important in treating disorders and diseases that may operate differently in men and women.

Clinton's find also piqued the interest of the egg industry, which would rather identify and destroy male embryos before they are hatched. While the sex of a chicken is determined at fertilization, it is extremely hard to tell the difference between the sexes of newly born chicks. Their color, size, and shape are virtually identical. Chicken sexing is an arcane art pioneered in the 1920s by Japanese masters that requires great skill. The sexer gently squeezes the bird's anal cavity to see if there is a small bump inside that indicates a male, but it is far more difficult than this description implies. "Successful chicken-sexing is a little like trying to recall a name or a dream that, for the moment, escapes you," the scientist Lyall Watson noted when he visited an Osaka center that trains some of the world's best. "The harder you try, the less successful you are likely to be."

Detecting a male embryo at the very start of the embryo's development would save the egg industry money by doing away with sexers and provide more space in incubators for the desired females. It also would halt what many animal-rights activists consider one of the most abusive practices in the industry. In the United States alone, more than 200 million young roosters are killed each year. In many documented cases, they were tossed live into Dumpsters or fed into wood chippers.

For such sex detection to be economical, it must cost less than one cent an egg and take less than fifteen minutes. Eggs today are typically removed from their incubator twice during their three-week development, once to check if the egg is fertilized and then later to vaccinate it, and a sex test would have to take place in one of these two short periods. If perfected, this process would save billions of roosters from execution days after their birth. It would also underscore just how superfluous the rooster has become.

The crowing cock dominated barnyards and chicken symbolism for millennia, but the age of the hen is dawning. Yet amid the modern world's alarm clocks and growing demand for unfertilized eggs, the rooster still has important religious duties to perform.

9.

Feeding Babalu

These birds daily control our officers of state; these . . .
order or forbid battle formation . . . these hold supreme
empire over the empire of the world.

—Pliny the Elder, *Natural History*

The Indonesian island of Bali has a unique culture that emphasizes beauty and balance in every aspect of life. It is also one of the few places in the world where cockfights are by law religious acts. If a sacred blood sport seems an oxymoron, it is just one of many on this Hindu outpost in the world's most populous Muslim nation, an ancient culture ringed by crowded beaches and rowdy bars. A meat lover's paradise within a vegetarian religious tradition, here the rooster retains its status as a sacred creature.

My first Balinese cockfight doesn't feel particularly holy. "Wanna bet?" asks a thin villager with a grin. A couple dozen men are gathered around a makeshift cockpit on a dusty corner across from a temple in a village near the island's center. Two men are squatting in the center of the bamboo enclosure, roosters between their legs. I Dewa Windhu Sancaya, a scholar of traditional Balinese culture who accompanies me, translates the request. I hesitate. He tactfully suggests that I make a "donation" to the temple. I hand over fifty thousand Indonesian rupiah, indicating with a nod my bird of choice. It is a scrappy-looking animal that resembles a red jungle fowl, only slightly larger. Within one minute, my money—worth about five dollars—is in another man's pocket and a skinny man at the curb is plucking the dead bird for the winner to take home to cook. There is no priest or blessing or prayer.

Balinese religion is a dazzling combination of animist practices and Hindu beliefs leavened with a sprinkling of Buddhism. Its fertile soil and well-engineered rice fields generate wealth that now is supplemented by the island's booming tourist industry. Elaborate rituals are at the core of Balinese life. There is a day when everyone stays at home and remains silent and even the airport is closed. There is a celebration of the goddess of wisdom and learning when people make offerings to books and abstain from reading and deleting anything written. There is even a festival that honors domesticated animals that are recognized for their role in human survival and the workings of the universe. Pigs are gaily decorated, cows washed and dressed in human clothes, and chickens and dogs are fed special treats as their owners offer prayers for their welfare.

The day after I arrive on the island is the festival of sharp objects, including motor vehicles as well as knives. Dazzling flower arrangements drape motorcycles. A large ceramic bowl filled with a dozen bananas and other exotic tropical fruits, flowers, and a well-cooked split chicken breast surrounded by smoking incense sticks sits on the hood of a parked car. The Balinese universe is a riotous mix of demons, nature spirits, ancestors, and gods and goddesses all clamoring for attention. Humans communicate with this world—a sort of spiritual extended family—via ceremonies designed to address the needs of a particular spirit or deity. Maintaining harmony in this system is the paramount goal. To do so, sacrificial rites may be made to an individual person, a deity, an ancestor, a priest, or a demon. Each offering is a gift or an appeal.

Later that afternoon, Windhu Sancaya takes me to visit a high priest at a nearby temple. The Brahman comes out to greet us in his stone courtyard. A lean older man with a weathered brown face and kind eyes, he invites us to sit cross-legged on a marble cushion-strewn platform in the center of the courtyard. Ida Pedanda Made Manis is fifty years old and comes from the priestly caste that lives on the gifts of villagers. *Pedanda* means "bearer of the staff" and *Made Manis* can be loosely translated as "second-born sweetie pie." When I ask if he enjoys a good cockfight, he chuckles with glee. "To be involved in gambling of any kind will keep us from our spiritual goal of mastering our attachment to the senses," he explains, turning serious. Then a smile returns. "Well, when I was young I would have liked to bet, but I couldn't afford it."

When I ask why chicken is the sacrificial animal of choice, he pauses. No one has ever posed such a question to him, the priest says, so he must seek permission of the gods before responding. He closes his eyes. In the silence, a warm tropical breeze swirls through the open space. Then he murmurs to an assistant. Moments later, a woven palm leaf tray piled with flowers, a few Chinese coins with holes in the center, and smoking aromatic incense appears. The priest gestures at me. "Take a coin," prompts Winhu Sancaya. I tentatively reach out, choose one, then place it gingerly into the open right palm of the Brahman. He squeezes his palm shut and once again closes his eyes, murmurs a prayer, and goes silent. Then he looks into my suddenly anxious face. "This question comes from your heart," the priest finally says with evident satisfaction. "And not just from your work. You are free to ask."

Chickens are favored, the *pedanda* begins, because they scratch around in the earth and will eat anything they can find. This makes them suitable for feeding demons but not heavenly deities; for those, duck and other animals are required. Then he explains Tabuh Rah, the concept of a sacred cockfight. "*Rah* means blood," the high priest says, leaning his slim frame slightly forward. "*Tabuh* means to purify so that the Bhuta Kala don't create disruptions," he adds. Bhuta Kala are negative forces or evil spirits that bedevil humanity, causing physical disease, mental illness, and societal havoc. Spilled chicken blood provides nourishment for the Bhuta Kala, keeping evil forces in check.

Sacrifice is at the core of Balinese beliefs. Though they enjoy a reputation for gentleness and compassion, islanders immolated multiple young women on pyres as recently as a century ago. A Balinese king told one early Western visitor that as many as 140 women would be taken by the flames when he died. Human blood is still used in some rituals in villages of eastern Bali, where it is drawn using a dagger, sharpened rattan sticks, or thorny leaves. The focus today is on animal sacrifice, and there is no other society on earth that practices in such numbers and with such regularity.

Terrorist bombings at a packed nightclub killed two hundred locals and tourists in 2002. To redress the imbalance created by the carnage, Hindu priests in white-and-gold robes slaughtered scores of water buffaloes, monkeys, pigs, ducks, cows, and roosters, and then placed their heads on altars to purify the devastated site. A celebrant drank pig blood drawn from a dead animal's throat while, in a boat offshore, priests weighted down two calves and threw them overboard. Most of Bali's 3 million residents took part in similar ceremonies around the island, which is half the size of Hawaii's big island. That ritual paled in comparison with a once-in-a-century ceremony in 1979 in which more than fifty water buffaloes, their horns covered in gold, were laden with precious goods and a large stone tied around their necks and then drowned in the sea while thousands of other animals, including chickens and ducks, were sacrificed to feed the Bhuta Kala.

The word for cockfight in Balinese is *tajen*, derived from the term for a sharp knife. It has been practiced here for at least a thousand years. "Night and day, you should hold cockfights" in the temple vicinity, according to one stone inscription carved in Old Javanese in AD 1011, one of the oldest known inscriptions on the island. A slightly later one states that "if you should hold cockfights within the sacred precinct" you are not subject to certain taxes.

The *pedanda* explains that the cockfight itself is not sacred, which is why the contest that I witnessed was held across the street from the temple. The cockpit traditionally is an elaborate open-air building called a *wantilan* that dominates a village or it is simply an open bit of ground. "If *tajen* is carried out in a holy place, human greed and passion will be there too," he adds. "It involves human greed; it is about accumulating funds, about winning and losing." But with the correct outlook, one of detachment from such worldly concerns, "Tabuh Rah is there." The separation between cockfighting and Tabuh Rah still seems fuzzy to me. Windhu Sancaya and other scholars of the complicated world of Balinese culture assure me later that such subtleties and paradoxes are rife in their ritual universe. The constantly changing world requires constant rebalancing, and Balinese rituals and beliefs must adapt accordingly. The Balinese universe would go awry without chickens. Few rituals of note take place without the death of the bird; only some die in a cockfight.

Dusk is falling quickly when we leave the *pedanda*'s compound. Windhu Sancaya takes me just down the road to a large religious complex where villagers are participating in an annual four-day celebration of the temple's founding. Festive coral and pink clouds, lit by the setting sun, hover above the stone courtyards, elegant pavilions, and tiered pagodas shaded by high-branched trees. Pura Penataran Agung Taman Bali, which means the great temple of the Balinese garden, resonates with the sounds made by the gamelan orchestra that plays under a stone-roofed arcade. Xylophones, drums, gongs, strings, and bamboo flutes bang and shimmer in the evening breeze. The atmosphere resembles that of a Midwest church picnic. Children play with red laser lights and SpongeBob balloons on the grass as women in gorgeous silks prepare a buffet dinner.

After the meal, there are prayers and a series of processions to honor the upper, middle, and lower worlds. Women carry massive offering bowls crowded with flowers and fruits as priests chant. After dark, the hundred or so participants troop down the worn stairs to the small courtyard at the temple entrance. The rising moon washes the stone in milky light. Dogs bark and bells ring, and incense smoke wreathes its way into the tropical sky. When everyone has gathered, the chatting stops and drums and pipes begin to play.

A clutch of a dozen young girls dressed in long gowns— premenstrual virgins, Windhu Sancaya explains—slowly begin to circle around an enormous oval of fruit and flower offerings laid in intricately shaped palm-leaf containers. The girls' stylized, birdlike dance reminds me of the female dancing spirits called *asparas* carved on the stone friezes of Cambodia's Angkor Wat temples. Attendants twirl huge black-and-white parasols. As one seated priest chants, another holds a chicken in his right hand and a bell in his left that he rings rhythmically. He takes up a short knife, slits the animal's throat, and pours the gushing blood into white bowls that are then emptied over the offerings. The two men kneel beside the mass of fruit and flowers and blood and play a game of egg toss until two eggs smash against each other and break. Everyone cheers.

The ritual complete, the *asparas* turn back into giggling preteens, mothers collect their tired children, and the men light up cigarettes. As the crowd disperses, Windhu Sancaya introduces me to two smiling priests. One wears brass buttons and a gold badge on his shirt like that of a Western sheriff and another is elderly with no obvious teeth. The chicken's blood, explains the one with the sheriff's badge, will feed the bodyguards of the gods who were invited from heaven to take part at the start of the ceremonies four days ago. These troops are earth demons that feed on blood and therefore must be satiated at the start and conclusion of the ritual.

The bird's blood is considered *rajas*—that is, endowed with activity and movement—while pig's blood, for example, is associated with inertia. Demonic bodyguards of heavenly deities, naturally, want the power of activity and movement that comes from the chicken's veins. The blood may be poured on offerings or used to mark crosses at cardinal points of a temple's boundaries. When I press him on why the chicken must be the source of blood, he says that it must be an animal with a close relationship to humans, part of our everyday life. "We have to sacrifice something we love," the sheriff says with a laugh, as if stating the obvious. "If you don't love it, then it is not a sacrifice. And the chicken is a symbol of the human family, because it comes in all colors."

White, red, and black chickens are much in demand for Balinese rituals, since each color represents a direction. But in the constantly shifting currents of Balinese theology, they also have come to represent the three human races. The killing is not simply to feed hungry demons. "Before we were human," the priest continues, "we were animals. And we hope that through sacrifice the animals can become human in their next incarnation." But along with this classic Hindu explanation, he offers another reason. "Long ago, humans were sacrificed," the priest adds. "Now, every blood sacrifice is done so that we don't have to use human blood."

This shift from human to animal sacrifice took place in many cultures. Ancient Chinese and Romans made the switch. In the biblical story of Abraham and Isaac, God orders Abraham to make a ram rather than his son the victim. But not just any creature will do. "We appear to substitute animals that are somewhat close to our social worlds," says Yancey Orr, a young American anthropologist at Australia's University of Queensland who studies Balinese culture. In the Abrahamic traditions, for example, God enjoyed Abel's gift of lamb blood but rejected Cain's hard-won offerings of vegetables and grain. They are too unlike and distant for human beings to use as a credible substitute.

Indonesia outlawed cockfighting in the 1980s to limit gambling and encourage sobriety and productivity in its citizens. Hindu Balinese were granted an exception. Technically, only three matches are allowed at a time and all gambling on birds is prohibited. I visited dozens of cockfights, large and small, around the island, where Balinese men delight in regularly breaking those rules.

But if their devotion to their game fowl is any measure, they sacrifice creatures that they dearly love. Men cradle their birds like children and fuss over their food and housing. "To anyone who has been in Bali any length of time," writes the late anthropologist Clifford Geertz, "the deep psychological identification of Balinese men with their cocks is unmistakable." The word for cock, *sabung*, can also mean anything from warrior or champion to lady-killer and tough guy. It is always a compliment.

It could be a Brooklyn block party, complete with a stage, concession stand, and milling late-night hipster crowd in a street closed off to cars by police barriers. But the stage on this balmy September evening is actually a two-story wall of blue and yellow plastic poultry crates. The concession stand is a slaughtering booth manned by expert butchers in blood-soaked yellow slickers and boots. The crowd is made up mostly of men with pale complexions above black beards and below black hats. They are waving live chickens over their heads.

It is nearing midnight on the eve of Yom Kippur, and the throng is here not to party but to expiate their sins. Here at the corner of Kingston Avenue and President Street, in the gritty neighborhood of Crown Heights, hundreds of Hasidic Jews are playing out a thousand-year-old ritual. At the center of this ceremony of *kapparot*, or atonement, is the pale-yellow bird. It will fortify their spiritual health for the Jewish New Year. The ceremony's ideal time is the hours before dawn of the day that marks Yom Kippur's start. In the quiet and dark of night, a calm that Hasidim call "divine kindness" is more easily accessible. On this night, here and in Hasidic communities around the world, the chicken's power to heal the spirit as well as the body is as alive as the dazed birds blinking in the spotlights set up on the sidewalk that illuminate the crowd.

Participants pay twelve dollars for a ticket and walk over to the wall of chicken crates that towers above the pavement, housing thousands of birds. Men get a rooster and women a hen. A pregnant woman buys three birds, two hens and a rooster, to cover herself as well as a baby girl or boy. Most hold their bird uneasily in their left hands, grasping the area between the wings, while balancing a prayer book in their right. Many young fathers conduct the ritual for their nervous and giggly sons. "Children of men who sit in darkness," the Hebrew chant begins, recalling the bird's ancient association with light. The worshipper waves the chicken around his or her head while saying, "This is my exchange, this is my substitute, this is my expiation." Each person repeats this act three times, for a total of nine chicken circles. Then the worshipper acknowledges that while the cock or hen is about to die, "I shall enjoy a long and pleasant life of peace."

Once that is done, the overwhelmingly male congregants take their chickens to the slaughter line. The booth is brightly lit and staffed by two burly men. They grab each proffered bird and swiftly and expertly slit its throat using a long and extraordinarily sharp knife. By kosher law, death must be swift and with a minimum of suffering. If the knife becomes dull and the animal is denied a quick demise, the bird is no longer considered kosher. Once butchered, the men toss the birds behind them as assistants in bloody yellow slickers scramble to shove the carcasses into large green plastic bags. Each bag then goes into a plastic trash can and is dragged to a van parked a few yards away. From there, they are taken to a soup kitchen; by tradition, the birds go to feed the poor.

Chickens are not mentioned in the Hebrew Bible and therefore are neither proscribed nor permitted as food. This posed a theological problem when the bird appeared in the Middle East. The Roman Jewish writer Josephus says that early rabbis were divided over whether the bird was kosher or unclean. Some scholars believe it was accepted and eaten in Galilee in the north, but forbidden in the sacred precincts of Jerusalem. At least one rooster lived close enough to the temple that the apostle Peter could hear it crow on the morning of Jesus's death. "Jerusalem, Jerusalem, you who kill the prophets and stone those sent to you," Jesus says in the gospel of Matthew, perhaps reflecting an upbringing among Galilean birds. "How often I have longed to gather your children together, as a hen gathers her chicks under her wings, and you were not willing."

The Mishnah, an older part of the Talmud compiled around AD 200, refers to chickens as *tarnegol*, or king's bird, a word derived from the ancient Akkadian term that reflects its earlier royal origins as an elite and exotic gift. A later portion of the Talmud praises the chicken as "the best of birds." The Hebrew word for rooster, *gever*, is the same as that for man, a curious fact that bolsters its status. The practice of *kapparot*, however, is mentioned neither in the Torah nor the Talmud, and has been controversial since it was first described in the ninth century AD by Jewish scholars at the Sura Academy south of Babylon, in today's Iraq.

A nineteenth-century historian reports that it was "a custom of the Persian Jews at an early date." That points to a ritual originating among the cock-adoring Zoroastrians who dominated today's Iran before Islam arrived in the seventh century AD. Mystics and common people embraced the ritual, while more bookish sorts tended to abhor it. Thirteenth-century rabbis dismissed it as a "foolish custom" derived from pagans, while a member of the Israeli parliament recently criticized it as "deplorable." The practice never gained many adherents in medieval Egypt and Spain. A small minority of orthodox Jews practices it today. Many of those are in New York City, and tens of thousands of birds are trucked in on the day before Yom Kippur.

The late-night Crown Heights gathering is rife with rabbis and scholars, and I ask several in the crowd to explain why the chicken is the favored animal. Rabbi Beryl Epstein, a Chattanooga-born Hasid with a ZZ Top beard, tells me that you could swing a potted seedling, a fish with fins and scales, or a white cloth filled with money destined for charity and achieve the same results. Like almost all the other pedestrians on the crowded sidewalk, he is wearing the high black hat and long black coat popular in eighteenth-century Poland where the mystical Hasidic movement began. "Here everyone does a chicken," he adds. Though swinging money is gaining popularity among other Hasidim, Lubavitcher Hasidim like Epstein prefer sticking to tradition.

Another rabbi tells me that since roosters and men share the same name, they are a good substitute for humans. A white-bearded scholar disagrees. He explains that any undomesticated four-legged animal—such as a deer—would do as well, but that chickens are simply easier to come by in New York City than wildlife. Yet another insists that it was the fowl's very absence from the Jerusalem temple that makes it the animal of choice. The ritual is not technically a sacrifice, since it involves use of a live animal. The killing part comes after the ritual. This matters to traditional Jews, since they are forbidden by rabbinical law from offering sacrifices in the wake of the temple's destruction in AD 70. Since the chicken never was a sacrificial animal, there is less chance that *kapparot* practitioners will confuse it with a sacrifice.

Opinions also vary within the crowd as to what is actually taking place during the *kapparot* ritual. Epstein says that the bird is not absorbing the sins of the person twirling it over his head. That would make repentance unnecessary. Instead, he sees the purpose as a "wake up" to the coming day of judgment, a reference that echoes the rooster's ancient role as a spiritual alarm clock. Other rabbis believe that the bird symbolically takes on the sins of the human. The fact that a chicken can't be "reused" for another person's *kapparot* suggests that the chicken is playing more than a symbolic role. Part of the prayer recited in the ritual includes the biblical verse that says God ordered people to be healed.

Most *kapparot* critics in the past focused on the pagan origins of the ritual, but animal cruelty is at the center of today's debate. In 2005, a chicken vendor abandoned more than three hundred birds packed into crates in a vacant lot in Brooklyn, drawing the attention of People for the Ethical Treatment of Animals. "The industrialization of *kapparot* has made it almost impossible to do the ritual humanely," PETA investigator Philip Schein told one reporter. "These are massive makeshift slaughterhouses on urban streets with tens of thousands of panicked chickens being trucked in horrible conditions and handled roughly by the public."

Lubavitchers like Epstein insist that the chickens are humanely treated, as required by Jewish law. But as I watch these city folk handle the live birds, it is clear that most don't know how to grasp the creatures without causing potential injuries. One teenage girl panics and flees when her chicken flaps its wings vigorously. Later, as I wander through the crowd, a few of the younger men confront me in halting English, questioning why I am taking pictures or ordering me not to. There's hostility in their voices, and I keep moving to avoid their angry stares. *Kapparot's* critics now include many orthodox Jews, and Lubavitchers are increasingly isolated in their practice. There is a new app for Yom Kippur atonement, but it uses a digital goat rather than the fowl.

Across the East River in Manhattan, a billboard showing a young Hasidic man gently holding a plump white chicken in his arms touts the "Alliance to End Chickens as Kaporos." A similar debate is being waged in Israel, as opponents to the practice become more outspoken. "Kapparot is not consistent with Jewish teachings," former Israeli chief rabbi Shlomo Goren said in 2006. The day before my visit to Crown Heights, the chief rabbinate declared that the use of the chicken was still permitted, but only if unnecessary suffering is avoided. Only later do I hear what took place not far from Crown Heights in the Borough Park section of Brooklyn. Hundreds of fowl slated for use in the ritual died from the heat of the October warm spell, *New York Daily News* reported—"No One Here but Us (Dead) Chickens!" was the eye-catching headline—and many people had to be turned away for lack of live birds.

Crown Heights' *kapparot* may be the last link with ancient traditions that used chicken sacrifice as a form of protective shield for children and a fertility ritual. In medieval times, the Jewish practice was designed primarily for children rather than adults. As recently as a century ago, Muslim villagers in Syria sacrificed chickens to ensure their progeny would survive and thrive, a hen for a daughter and a cock for a son. This tradition even extends to the other side of the world, in the Babar archipelago in eastern Indonesia. James Frazer noted in his 1890 book *The Golden Bough: A Study in Magic and Religion* that a Babar woman desiring a child would have a man hold a chicken over a woman's head and repeat the words: "O Upulero, make use of this fowl; let fall, let descend a child, I beseech you, I entreat you, let a child fall and descend into my hands and onto my lap." He then holds the bird over the man's head, says another prayer, and kills the chicken.

The healing power of chickens in a religious context was common even in nineteenth-century Wales. Frazer notes that epileptics would come to the village church of Llandegla to perform a ritual in which their malady was magically transferred to a rooster or hen, depending on the patient's sex. Jews in nineteenth-century Galicia believed that epilepsy could be cured using a slaughtered cock. Such traditions may seem bizarre, but they are remnants of a time when the chicken was an elemental part of our spiritual as well as our physical well-being. One surviving bastion of the bird's spiritual healing power is in a Miami suburb.

The morning after Arkansas governor Bill Clinton won the 1992 presidential election, Justice Antonin Scalia discussed chicken sacrifice on the bench of the U.S. Supreme Court. "You may kill animals for food but not for other purposes?" he asked the lawyer arguing for repeal of a Florida law forbidding the practice. "Not for sport, not for sacrifice, not for anything but food?" The lawyer responded that there might be one exception. "You'd kill an animal in self-defense if you're being attacked by a bear."

This surreal exchange took place during oral arguments in a landmark battle over religious freedom. What brought the case to the attention of the country's highest court was the ritual slaughter of chickens. The trouble began when Ernesto Pichardo decided to build a church. A priest in a congregation practicing Santeria, an African-Caribbean religious tradition popular in Cuba, Pichardo sought permits from the city of Hialeah in southern Florida to create a small sanctuary and cultural center in the overwhelmingly Hispanic and Catholic community.

Starting in the late 1980s, fears of satanic rites conducted in secret in which animals and even children were sacrificed were sweeping the country. "Blood Cults Spread through U.S.," warned one newspaper headline. Santerian refugees from Cuba were caught up in the panic. Drawing on ancient African practices, Santerians regularly sacrifice chickens and occasionally goats in private rituals to feed primary spirits, called orishas, sometimes equated with Catholic saints. Santeria means "the way of the saints."

Many Christians in Hialeah equated Santeria instead with devil worship. City officials reacted with horror, disgust, and outrage to Pichardo's request. "What can we do to prevent this church from opening?" the president of the city council bluntly asked. What they could and did do was pass laws that made animal sacrifice a crime, unless it was primarily for food consumption. During an emergency public session in 1987, residents and city officials taunted Pichardo when he spoke against the ordinances and cheered when his opponents criticized Santeria. "People were put in jail for practicing this religion" in pre-Castro Cuba, asserted one hostile city council member, to the applause of many in the room.

Pichardo's Church of the Lukumi Babalu Aye took the case to court, arguing that the laws were unconstitutional. Two lower courts backed the city of Hialeah's arguments that the community wanted to prevent animal cruelty, limit the spread of disease, and shield children from the trauma of witnessing bloody killings. So in late 1992, Pichardo and his lawyers took the case to the land's highest court.

Lukumi is another term for Santeria, the religion brought to the United States from Cuba by refugees like Pichardo in the aftermath of Fidel Castro's revolution in 1959. The Mariel boatlift of 1980 brought more than 125,000 Cubans to the United States, most to south Florida. Many practiced Santeria, and a tradition that had existed only on the margins of American life began to raise its profile just as fears of sacrificial cults spiked.

Most Americans already were familiar with one of the principal orishas of Santeria. "Babalu," Ricky Ricardo's theme song in the 1950s television show *I Love Lucy*, is the earth deity associated with illness and healing, sometimes tied to the Catholic saint Lazarus. White hens and roosters are often sacrificed in his name, while black ones are used to absorb an unwanted spell. "Hens and roosters are sensitive to evil forces," one Santeria vendor at a stall in Caracas, Venezuela, explained to me.

Santeria's roots stretch across the Atlantic to what is today Nigeria, which faces the Atlantic coast in West Africa. About AD 1200, a settlement called Ile-Ife emerged as an important artistic, intellectual, and religious center in West Africa for the next two centuries. Even today it is considered the Jerusalem of the 35 million Yoruba people who make up one of Africa's largest ethnic groups.

According to one Yoruba tradition, the earth itself began at Ile-Ife.

The creator deity Oduduwa dropped a chain from heaven but found only water below. The god then emptied a basket of earth and placed a five-toed chicken on this little mound. By scratching the soil, the chicken spread around the dirt and made more dry land. "Where the claws dug deep, valleys were formed," one version relates. "Hills, uplands, and mountains were left behind within the interstices of the claws." Then Oduduwa planted a palm nut that grew into a sacred tree.

This is the site of Ile-Ife. The words can be translated as "holy life" or as "earth spreader." "Such a name for the most ancient Yoruba kingdom suggests that it may have been named for the giant chicken that provided the earthly foundation on which it rests," writes Daniel McCall, a Boston University anthropologist who did fieldwork across West Africa. Yoruba art is full of wooden sculptures of kneeling individuals offering up large chickens. Often the chickens are in the shape of a bowl for storing palm nuts. McCall suspects that some of these depict Oduduwa and his formidable fowl. Medieval Yoruba priests called *babalawos* frequently used chickens for divining the future, curing illness, and dispelling evil forces.

Chickens came late to sub-Saharan Africa. The first mention of the bird in this vast region comes from the Arab adventurer and pilgrim Ibn Battutah, who traveled in what is Mali in 1353, trading with village women for live fowl. There was no archaeological evidence for earlier African chickens until 1991, when the British archaeologist Kevin MacDonald found a bone in Mali dating to five centuries before Battutah's visit. Researchers are now trying to piece together the story of the bird's spread in this region, which coincided with the rise of sub-Sahara's first complex societies. In a 2011 Ethiopian dig, excavators found a fourth-century BC bone, suggesting that chickens first arrived from Egypt or by boat across the narrow strait separating the Horn of Africa from south Arabia, perhaps accompanying a cargo of aromatic frankincense and myrrh. Later imports were part of the bustling Indian Ocean trade in early medieval times, when Arab, Indian, and Indonesian merchants plied the waters of Africa's east coast. Genetic analysis of Nigerian chickens links them to distant Southeast Asian rather than Mediterranean varieties.

Facing stiff local competition, however, the bird was slow to spread in sub-Saharan Africa. Pigeons and francolin were abundant, and at least one variety of guinea fowl was already domesticated. Chickens don't thrive in open grassland or stony desert and they are vulnerable to predators in the jungle, so they were of marginal importance to the peoples who live on Africa's savannahs or in dense forests. But the birds were tailor-made for the farming communities and towns that began to flourish in West Africa in early medieval times. They ate noxious insects, provided abundant eggs and meat, and had a social nature that fit in well with barnyard life.

The chicken's arrival in West Africa was nothing less than revolutionary, at least in Kirikongo. Stephen Dueppen, a professor at the University of Oregon in Eugene, has been working since 2004 at this site in the savannah of Burkina Faso, between the Sahara Desert to the north and the dense jungles of Ghana to the south. A lean thirtysomething with a Los Angeles accent, Dueppen is exploring a rare and remarkable time capsule of an ancient African village that grew from a single homestead around AD 100 into a small settlement abandoned in the seventeenth century.

He excavated a chicken bone dating to AD 650, the oldest yet found in West Africa. By that date, the cattle-herding Kirikongons had a growing elite, and by AD 1000 the leading family buried its children with luxury goods like cowrie shells brought from the distant ocean. It's the familiar story of civilization—elites develop in small villages, long-distance trade takes off, towns appear with political and religious leaders, and then cities, kings, and empires emerge. But about AD 1200, the villagers of Kirikongo abruptly strayed off message.

"It's a pretty dramatic change," says Dueppen. He's invited me to visit his excavation, but unrest in neighboring Mali and a rash of kidnappings near its border force him to cancel the season. Suddenly, people are pounding their grain in public spaces rather than in isolated compounds, villagers have paved over the cemetery, and all the cattle are gone. In a region where cattle are still considered a primary source of wealth and prestige, this last decision was momentous. The large animals carry religious power. Even today across much of the savannah, a man's standing in the community—and his marriage opportunities—often hinge on his herd size. Cattle require large grazing areas and a high level of organization. Just as Mongolian herders like Genghis Khan expanded their range, empires in this part of Africa grew out of a culture of large livestock. Since there is no evidence of sudden climate change or a devastating cattle disease, the people of Kirikongo appear to have made a conscious decision to create a different sort of society.

As these changes were taking place, the village built a ritual complex, possibly with an attached communal granary. The floor of the entire structure is paved in brick, but there is one small, unpaved section. Here, Dueppen's team uncovered the remains of at least four chickens, a goat, and a stone for sharpening knives. "You want the blood of the sacrifice to drip into the dirt, as a donation to the ancestors," he explains. This is, at least, the current tradition, and what he found in the ancient mound is similar to the altars one might find in a nearby village even now.

Dueppen believes that the rejection of cattle and hierarchy hinged on the availability of the chicken. Come the revolution, the fowl provided a viable alternative to the cow. Chicken bones quickly replace those from cattle. Cattle are traditionally a male domain, whether in Britain or Burma, while chickens are typically under female control in pre-industrial cultures. Dueppen suspects that women may have played a key role in overturning Kirikongo's nascent elite.

When the villagers broke the power of the elite they also made it possible for more people to participate in sacrifices. Until then, these rituals revolved around expensive cattle owned by a few. This shift increased the demand for chickens, or the availability of chickens increased the opportunities for sacrifice. Among the local Bwa people who inhabit the area today, the bird is used on virtually any important occasion. Chickens, the archaeologist explains, enable prayer. Any ritual involving ancestors, politics, divination, judicial proceedings, as well as births, marriages, and deaths all involve sacrificing the bird. In the past, even iron could not be smelted without a proper chicken offering. Among neighboring Mali Sudanese, "to sacrifice this bird is to sacrifice a substitute for the world."

The bird made it possible to make offerings without straining a family's budget. In his 2010 paper in *American Antiquity*, Dueppen writes "chickens provided a means to maintain the rich spiritual life within a consciously reinvented egalitarian society without creating wealth differentials." In other words, you didn't need to be a rich cattle baron to maintain your connection with the gods. Kirikongo's past, he maintains, is a unique case study of "the reinvention of equality in the creation of a novel society."

The chicken revolution still reverberates in the region. Men can marry without selling off their land or animals and women can divorce their husbands with a freedom not found in many other West African societies. The egalitarian nature of the modern people living in this area extends to gender. If your goal is to build a more egalitarian society, then chickens are the perfect livestock. "People forget about chickens because they are so common," Dueppen says. "But they have a bunch of advantages that other animals do not. They are low cost and quickly reproduce. They are omnipresent. They are just so flexible. And they work in any political system."

The bird assumed a pivotal role in many African religious traditions. In the Congo basin, a female shaman among the Lulua tribe undergoes a rigorous ordeal of death and rebirth, but is not considered properly initiated until a hen is placed around her neck with the power to "lure the souls of dead mediums" back to the earth. In Sierra Leone, on the Atlantic coast, the chicken is a truth teller. If it pecks at the grain in the hand of an estranged relative or friend, then the dispute is resolved. Among the Ndembu in central Africa, a chicken determines the guilt of a person accused of committing a crime. Poison is placed in a chicken's mouth. If it survives, a dose is given to another chicken. If that bird lives, poison is administered to a third bird. Only then, if the third lives, is the accused found not guilty.

Among the farming and town-dwelling Yoruba, south of Burkina Faso's Kirikongo, the foreign chicken eventually displaced the indigenous pigeon as the center of their creation mythology. Yoruba proverbs are liberally sprinkled with references to chickens—"Were the hawk to die, the owner of chickens would not shed a tear." The traditional Yoruba universe is made up of a heaven crowded with ancestors and hundreds of orishas, and an earth below filled with people and animals. As in so many other disparate cultures—American Hasidim and Hindu Balinese, for example—the chicken serves as a tool or a medium to connect us to heavenly healing. Its blood seals that connection. Many Yoruba deities require its sacrifice. Births, weddings, illnesses, and deaths were, and often still are, accompanied by chicken sacrifices.

It is worth recalling that chicken sacrifice and divination were widely practiced among Persians, Greeks, Celts, and Germanic tribes in the West, and Southeast Asians and Chinese in the East. But no society brought chicken divination to the level of organization, complexity, and importance than did the ancient Romans. Today's Santerian practices could have been conducted in ancient Rome without provoking comment, much less legal action. Chicken sacrifice and divination was not only expected, it was critical for public policy. No major decision about war or peace was made without such rites.

The Roman traditions still survive in our words. When we refer to an auspicious occasion, we are harkening back to this tradition. The term *auspice* comes from Latin, and derives from "observer of birds." Augury—that is, divining the will of the gods by observing the natural world—is thought to come from the Latin root words that mean "directing the birds," though it more likely stems from *aug*, meaning to prosper. Whatever the word's origin, an augur was a state-sponsored Roman priest seeking to know the gods' plan largely through watching birds. This was a serious job reserved for aristocratic men of dignity, notes the famous first-century BC philosopher and orator Cicero.

Cicero was an augur himself. Before any major undertaking by the state, specialists scanned the skies to mark the behavior of eagles, ravens, or other creatures that flew in the heavens. Their movements, if properly interpreted, could reveal whether the gods favored a particular course of action such as starting a war or suing for peace. After all, in this era only birds shared the heavens with the gods.

By Cicero's time, in the waning days of the Roman Republic, cynical politicians had hijacked the hallowed tradition of augury. "Scarcely any matter out of the ordinary was undertaken, even in private life, without first consulting the auspices," he writes. As the empire grew, however, watching for hawks or eagles or pigeons in the skies above the capital proved too haphazard. Wild birds come and go depending on weather, time of day, and migrating habits. They didn't always appear when their assistance was required.

Domesticated chickens, by contrast, could be kept close by for frequent consultations at any time of night or day, so the Roman government gradually shifted its focus to the fowl. A specialist called a *pullarius* cared for the sacred flocks, which were kept in temples near the forum, among legions, and even on ships. The primary means of extracting an omen from these birds was for the pullarius to toss grain, bread, or cake into the coop. If the birds gobbled up what was proffered, the action in question had the gods' blessing. But if they refused to eat or, worse, made a racket and moved away from the food, this portended ill.

In Cicero's day, people still recalled what took place on the morning of a crucial naval battle two centuries before. The sacred chickens kept on a Roman warship refused to eat the grain, which wasn't the omen desired by the arrogant consul, who had the offending birds tossed overboard, allegedly saying, "Let them drink, since they don't wish to eat." The enemy defeated the Romans. His blasphemy against the sacred creatures was not forgiven or forgotten.

A senior Roman general likewise scoffed when an augur informed him that the chickens' omens required that he remain in camp rather than fight. He and most of his army were slain within three hours as devastating earthquakes shook Italy. "A large number of towns were destroyed . . . the earth sank, rivers flowed upstream and the sea invaded their channels," Cicero notes. Contradicting a chicken in ancient Rome was no laughing matter.

This reliance on coop-bound birds naturally made the omens easier to manipulate. Chickens could be starved or overfed in order to produce the desired results. Cicero, Rome's consummate statesman who would be assassinated for his opposition to dictatorship, had few illusions about how religious rituals could be twisted by self-serving lawmakers. "I think that, although in the beginning augural law was established from a belief in divination, later it was maintained and preserved from considerations of political expediency."

Our word *sacrifice* comes from Latin, meaning to make sacred. Chickens, along with other animals like pigeons or cows, were regularly sacrificed in Rome on both private and public occasions. Sometimes specialists called *haruspices* would then examine their interior organs, which provided data on whatever issue—illness, birth, money problems—was at stake for a person, a family, or the nation. The offerings then typically were cooked and consumed by priests and the laity taking part in the ritual. The echoes of this ancient practice are still found in Jewish kosher and Muslim halal rules. These laws are based on the knowledge that humans kill to live, and a set of methods and prayers are necessary to maintain a respectful relationship with the deities that make our lives possible.

Now, of course, slaughter and evisceration of the animals we eat increasingly takes place out of the sight of today's urban dwellers, removed from our daily lives. Even most modern kosher and halal killings are conducted in factory plants, and the prayers are often pre-recorded and played on an endless loop. By arguing for their right to conduct ritual sacrifice, Pichardo and his church prompted a national debate about animal cruelty, witchcraft, and religious freedom.

"In a world of instant gratification such as our own, the concept of sacrifice might seem strange," writes Ócha'ni Lele about Santeria practices. "But in truth, in every moment of life sacrifices must be made for the betterment of oneself and one's community." To the Lukumi, sacrifice is at the very heart of the faith that came on the slave ships that crossed the Atlantic.

All African slaves brought their particular traditions to the Americas. But they were often split up from their families and ethnic groups, weakening traditional links. Fearful of revolts and witchcraft, white owners and governments did their best to stamp out practices seen as pagan and substitute mainstream Christianity in its place.

Some practices survived in places like the U.S. South, but in Cuba whole traditions remained largely intact. In the early nineteenth century, thousands of Yoruba were sent to this Spanish-run island dominated by sugar plantations. Their late arrival, large numbers, and concentration around Havana helped protect Yoruban traditions despite laws banning many of their practices. Under a semi-Catholic facade, orishas transformed into saints. Babalu Aye became Saint Lazarus, and chickens continued to be sacrificed, though often in secret to avoid punishment by wary whites.

Testifying to the U.S. district court, Pichardo explained that only priests conduct the sacrifices by puncturing the carotid arteries in one move, much as is required in Jewish and Muslim religious rules. The animal's blood then is drained into a clay pot and the head is decapitated. Like Rabbi Epstein in Brooklyn, Pichardo insists that the animals are well cared for and that their deaths are quick. Animal rights activists counter with numerous cases both among Hasidic Jews and Santerians where such humane intentions are ignored.

For Santerians there is the added complication that those chickens designated to absorb spells or illness from individuals cannot be eaten and must be left outside to decompose. Sanitary workers at the Miami-Dade County Courthouse must clear the steps of dead and decapitated chickens that were part of various rites, and joggers along the Miami River periodically must be careful of treading on rotting carcasses. For urban Americans, this can be deeply disturbing. But for many Cuban emigrants, these dead animals are a necessary part of life.

In June 1993, all nine Supreme Court justices, including the devoutly Catholic Scalia, strongly backed the right of Pichardo and his congregation to practice their sacred rites. Hialeahan officials "failed to perceive, or chose to ignore the fact that their official actions violated the Nation's essential commitment to religious freedom," scolds Justice Anthony Kennedy in the unanimous opinion. "The sacrifice of animals as part of religious rituals has ancient roots," he wrote. "Although the practice of animal sacrifice may seem abhorrent to some, 'religious beliefs need not be acceptable, logical, consistent, or comprehensible to others in order to merit First Amendment protection,'" Kennedy adds, quoting an earlier decision.

There were no birds sacrificed when Chief Justice William Rehnquist inaugurated Bill Clinton as president shortly before the Hialeah decision was issued, but ritual chicken killings in the United States are not confined to Miami and its suburbs. U.S. Park Police recently found the remains of sacrificed birds just a few miles from the high court in Washington's Rock Creek Park. Pichardo and his congregation went on to build their modest church and community center. Other Santeria sanctuaries have since sprung up around the country, and the fear and suspicion surrounding the tradition has largely abated. Ciccro might have been pleased.

10.

Sweater Girls of the Barnyard

My honest friend, will you take eggs for money?

—William Shakespeare, *The Winter's Tale*

Among African Americans across the South, the pastor had first dibs on the bird that typically graced the table after church. "He ate the biggest, brownest, and best parts of the chicken at every Sunday meal," complains poet Maya Angelou, recalling the pious glutton Reverend Howard Thomas. Often called the preacher's bird or the gospel fowl, echoing its sacred role among West Africans, slaves and their descendants laid the foundation for America's love affair with chicken that is now spreading around the world.

British settlers brought their flocks to Jamestown in 1607, a dozen years before the first African slaves arrived on Virginia's shores, and the birds helped sustain struggling colonists. Food grew so short in 1610 that the governor made unauthorized killing of any domesticated animal, including cocks and hens, punishable by death. In New England, where chickens arrived on the *Mayflower*, Edward Winslow in 1623 sent an ailing Native American chief two chickens to make a healing soup. Grateful for the exotic gift, the chief is said to have revealed a plot by another tribe to destroy the nascent colony.

The bird, however, was rarely more than an incidental food in colonial America. On Winslow's farm, wild bird remains excavated by archaeologists outnumber those of chickens three to one, and cattle, pig, sheep, and goat bones dominate. Virginians, meanwhile, feasted on turkey, goose, pigeon, partridge, and duck, along with venison, mutton, pork, and beef, as well as shad, sturgeon, and shellfish. "Seventeenth- and eighteenth-century descriptions of colonial foodways ignored the chicken for the most part," says the *Oxford Encyclopedia of Food and Drink in America.*

For enslaved African Americans, this humble status proved a welcome boon. In 1692, after several individuals bought their freedom with profits from animal sales, the Virginia General Assembly made it illegal for slaves to own horses, cattle, and pigs. Masters often forbade their human chattel from hunting, fishing, or growing tobacco. The chicken "is the only pleasure allowed to Negroes; they are not permitted to keep either ducks or geese or pigs," one visitor to George Washington's Mount Vernon remarked.

On the expanding farms of the colonial South, African Americans began to breed, buy, sell, and eat fowl as they saw fit. Slaves in this era typically were allowed to grow their own vegetables, and chickens were fed garden waste, table scraps, and the roughage from ground corn. When Washington ordered cornmeal to replace the whole kernels formerly given to his Mount Vernon slaves, they complained, "as much from the want of the husks to feed their fowls, as from any other cause," he wrote.

Chicken bones account for a full third of all bones found among the African-American quarters of Maryland and Virginia plantations, and plantation records show frequent cash payments to slaves for poultry. Owners granted slaves authority over chickens because the birds were of negligible economic importance and reduced plantation spending on feeding field hands, and many slaves of West African descent retained skills in poultry raising inherited from their ancestors. Virginia planter Landon Carter, for example, mentions two hundred chickens "entrusted" to his slave Sukey.

Just as European Jews gained expertise in money lending, a

profession disdained by Christians, poultry became an African-American specialty. Some planters required slaves to sell them all surplus meat and eggs in order to limit their entrepreneurial freedom. As early as 1665, Maryland governor Phillip Calvert sued Thomas and Elizabeth Wynne for buying ten chickens from his slaves, who had pocketed the proceeds. A century later, Washington's neighbor James Mercer wrote that African Americans "are the general chicken merchants." In one letter to his overseer he offered to exchange several yards of linen to "pay for some debt I am said to owe them for Chickens."

In what likely was a typical exchange, Thomas Jefferson in 1775 bought three chickens for two silver Spanish bits from two of his female slaves who worked at his Shadwell plantation. In the early 1800s, when he was away serving as president of the United States, Jefferson's granddaughter Ann Cary Randolph recorded each purchase made while she helped manage the estate in his absence. Chickens and eggs were the items most commonly sold by slaves to the white household; the only ones not involved with poultry sales were the cooks and Jefferson's mistress, Sally Hemings. Such commerce took place on South Carolina's rice plantations as well, where slaves often were expected to grow their own food. In 1728, a white owner named Elias Ball paid one pound and fifteen shillings for eighteen chickens to his slave Abraham. A savvy businessman, Abraham then threw in one chicken for free. Planters bought up to seventy birds at a time from their human chattel.

"Adjoining their little habitations, the slaves commonly have small gardens and yards for poultry, which are all their own property," wrote Isaac Weld while visiting Virginia from Britain in the 1790s. "Their gardens are generally found well stocked, and their flocks of poultry numerous." Free and enslaved African Americans created extensive chicken networks both on and off plantations, with free blacks serving as middlemen to distribute birds beyond the borders of plantations. One traveler passing through Virginia before the revolution was surprised "by a crowd of negroes" who showed up at his door eager to sell poultry and produce to the newcomer. In Charleston, black women at the city market sold chickens and eggs "from morn 'til night," the *South Carolina Gazette* reported. In that setting, they were free to charge their white customers high prices and keep the proceeds. "The slaves sell eggs and chickens," wrote Fredrika Bremer, an intrepid Swede who toured the antebellum South. "They often lay up money; and I heard speak of slaves who possess several hundred dollars."

The birds were also currency in the underground economy among African Americans. Chicken-rich slaves on one Virginia farm contracted enslaved carpenters to build wooden stools for their cabins in exchange for fowl. By the time of the Civil War, African Americans had been selling chickens and eggs for more than two centuries. This thriving business did not extend into the Deep South, where conditions were often harsher and entrepreneurship was more difficult. In one South Carolina rhyme from the era before the Civil War, masters threatened troublesome slaves with deportation to "Ole Miss'sip' whar de sun shines hot, Dere hain't no chickens an' de Niggers eats c'on."

Slaves involved in the poultry business had a strong economic incentive to encourage their masters to eat more chicken. Since black women often did the cooking in plantation kitchens, West African foods like okra and kale crept onto plantation menus. But it was an aborted slave rebellion that helped launch fried chicken as a favorite Southern dish among whites as well as blacks.

In 1800, as slaves fought a revolution against their French masters in Haiti, an enslaved Richmond blacksmith named Gabriel Prosser organized a revolt in the Virginia capital. The plan was discovered and he was jailed with his fellow conspirators. In a deposition, Prosser said that he had intended to make Mary Randolph, a well-known white Virginia hostess and chef, his queen in a new African-American-controlled regime. Her infuriated husband, Richmond's federal marshal, sought execution for all the plotters. Jefferson, then campaigning for the presidency, counseled Governor James Monroe to be lenient. Over Randolph's objections, Prosser and two dozen men were hanged, but ten others were spared.

The new Jefferson administration dismissed Randolph from his government post as tobacco prices collapsed. The couple lost much of their wealth, and Mary Randolph—known now as "Queen Molly"—opened a Richmond boardinghouse to make ends meet. Her fame as a chef spread and in 1824 she published *The Virginia Housewife*, considered the first Southern cookbook. Designed to replace reprints of old English cookbooks, Randolph's list of recipes combined British and African traditions using American ingredients, including the first published recipe for Southern fried chicken. She advised dredging pieces in flour, sprinkling them with salt, and frying them in lard to a light brown color.

Fried chicken is not unique to West Africa; one ancient Roman recipe calls for frying the bird with spices, including pepper, and Scottish Highland cooks have an old tradition that came with them to the New World of placing chicken pieces in an iron pot with hot oil. But Randolph's recipe, copied or heavily influenced by African-American slaves, became the template for the dish. A century later, another white, a Midwesterner named Harland Sanders, combined a variation of this recipe with the technological innovation of the pressure cooker to launch the world's second-most lucrative fast-food chain, Kentucky Fried Chicken. After three centuries of cultivating flocks, cooking fowl, and laying the foundation for the modern poultry industry, African Americans found themselves relegated to the sidelines and stereotyped as chicken thieves even as fried chicken emerged as a quintessentially American food.

A quarter century after Randolph published her cookbook, a virulent strain of hen fever jumped from Britain to New England. More than ten thousand people converged on a chilly Thursday morning in Boston's Public Garden in November 1849 as politicians, businessmen, bureaucrats, and the general public crowded the nation's first fancy poultry exhibition. "Everybody was there," wrote eyewitness George Burnham. Daniel Webster, the nation's greatest living orator, with a voice likened to a church organ, came with a pair of Javanese fowl. Urged to give a speech, his attempt was drowned out by the cacophony of roosters. One man that day paid thirteen dollars for a single pair of chickens, the price of two barrels of wheat flour.

Burnham was a gentleman scoundrel, newspaperman, and poultry breeder who helped stoke the American poultry fancy. A month after the Boston show, he purchased a half-dozen Cochin China fowl from Lord Heytesbury, the Lord Lieutenant of Ireland at the height of the potato famine. These were from the same stock that Queen Victoria and her prince consort, Albert, had sent from their poultry house at Windsor Castle for the Dublin show held that previous spring. He paid the enormous sum of ninety dollars—about twenty-five hundred in today's dollars.

The arrival in Boston of these birds, which a local paper deemed "extraordinary, and strikingly peculiar," created a sensation. Nine months later, Burnham received sixty-five dollars by selling just four young birds raised from the original fowl, proving that he had made a wise investment. In 1852, he sent a pair to Queen Victoria in a publicity stunt. In gratitude, the grateful monarch sent her portrait to Burnham, and the entire affair was covered in the newspapers. At a time when a factory worker earned an average of seven cents an hour, the price for a pair of Cochin fowl shot up from $150 to $700.

The consummate American showman, P. T. Barnum, fell under the spell of fancy chickens as well, building large coops on his lavish Iranistan estate in upstate New York and serving as president of the National Poultry Society. The abolitionist editor Horace Greeley and Burnham were vice presidents. Barnum organized the first national poultry show in February 1854, using the buzz about the birds as a way to pull people into his museum on Broadway in New York City. "There will be a marvelous cackling as the sun gets up this morning," noted the February 13 *New York Times*. Offering $500 in prizes—the equivalent of about $13,500 today—Barnum advertised "all foreign and just imported" chickens. The crowds prompted Barnum to extend the show a full six days. He repeated the wildly successful exhibition that October, when he had a composer create a "National Grand Poultry Show Polka." Like Burnham, the showman made a fortune from hen fever.

The big money, extensive publicity, and near hysteria surrounding hen fever alarmed the nation's sober-minded agricultural press. Chickens were minor players and associated with women in a male farm culture dominated by grain and large livestock. Editors warned the public against paying exorbitant sums for "Shanghaes, Chittagongs, Cochin-Chinas, Plymouth Rocks, and a half a dozen other puffed-up worthless breeds of fowls." Few heeded this advice. An upstate New York newspaper reported that one farmer who sold his Cochin China birds for $10 a pair and $4 for a dozen eggs made $433 in a year—a hefty fortune for a small farmer in the 1850s. New England clergy speculated heavily in Chittagongs and Brahmas, and white Southerners complained that their slaves had access to the new and expensive varieties. A plantation owner in Rome, Georgia, asserted in 1853 that his black workers raised five hundred Shanghai fowls the previous year, hefty-breasted chickens "on which they daily luxuriate."

The writer Herman Melville satirized hen fever at a time when ten Shanghais could cost six hundred dollars. In his 1853 short story "Cock-a-Doodle-Doo! or The Crowing of the Noble Cock Beneventano," his protagonist becomes so obsessed with a Shanghai rooster that he is willing to mortgage his farm to obtain it. He wrote:

"A cock, more like a golden eagle than a cock. A cock, more like a field marshal . . . Such a cock! He was of a haughty size, stood haughtily on his haughty legs. His colors were red, gold, and white. The red was on his crest alone, which was a mighty and symmetric crest, like unto Hector's helmet, as delineated on antique shields. His plumage was snowy, traced with gold. He walked in front of the shanty, like a peer of the realm; his crest lifted, his chest heaved out, his embroidered trappings flashing in the light. His pace was wonderful. He looked like some Oriental king in some magnificent Italian opera."

As it did in Britain, the bubble burst in 1855. Like the more famous California Gold Rush that boomed and busted in the same years, hen mania was short in duration but had dramatic long-term effects. By the time the Civil War broke out, new varieties of chickens peppered American farms on both sides of the Mason-Dixon Line.

Collectors who sought fowl with dramatic colors and outrageous feathering began to experiment with breeding meatier birds that laid more and larger eggs. When the bloody conflict was over, American enthusiasts were poised to create the chickens that now account for the vast majority of the world's poultry.

The Plymouth Rock, one of the most common modern varieties, was first shown at the 1849 Boston show by John Cook Bennett, a onetime leader of the Mormon movement who advocated free love and launched the use of the anesthetic chloroform. His new variety combined genes of English Dorkings with the Asian Cochins and Malays, and one poultry fancier dismissed it as "a mongrel of little worth." Later versions mixed the genes of Dominiques with Cochins or Javas to create a meaty bird that also was a reliable egg layer.

In 1875, a Maine farmer bred the first white Plymouth Rock, a bird descended from Asian and European varieties that would eventually spread worldwide. Rhode Island Reds, meanwhile, came from crossing Asian varieties with the Leghorn, a bird imported from Italy in the 1840s. British-bred crosses like the Cornish hen and Orpington also gained popularity. Some new breeds specialized in churning out eggs or increasing meat productivity.

Railroads, mines, and manufacturing plants spread across the North as the Industrial Revolution gathered steam. Urban populations ballooned and food demand jumped while traditional ways of farming gave way to new industries such as the cattle business. By the 1880s, beef from Texas cattle slaughtered in Chicago ended up on plates in New York, London, Paris, and Berlin. The complex system of cattle drives, rail shipment, slaughterhouses, and steamship transportation helped ensure that beef would remain Americans' meat of preference for a century. Pork processors did their best to imitate the cattle system.

Chicken was different. The bird required only a protective coop and could survive on kitchen and garden scraps. Free feed and labor—usually performed by farm women—made it difficult to imagine poultry on an industrial scale. Some nineteenth-century Americans turned to ancient Romans for advice. While holding the bird sacred, Romans also loved roast chicken. In the first century AD, the scholar and farmer Marcus Terentius Varro had detailed advice for "one who wants to set up a poultry-farm . . . to gain a large profit by the exercise of knowledge and care." Varro recommended housing a flock of two hundred birds indoors. Peacocks and guinea fowl could also be kept, but "it is chiefly the barnyard fowls which are fattened." These chickens, he adds, "are shut into a warm, narrow, darkened place, because movement on their part and light free them from the slavery of fat." They might be sacred, but they were not free range.

Writers like Pliny the Elder and Columella discuss in great detail the rate of egg production, the advantages of certain breeds, the diseases that afflict chickens, and methods to house, feed, and protect them. Pliny kept his chickens in tiny pens, which we would today call battery cages, so that they would continually eat. "We disregard the chief purpose of the Greeks, which was to train all the fiercest birds for fighting," explains Columella, who expresses a patrician Roman's disapproval of gambling. "Our object is rather to bring in an income to the hard-working family man, and not to the trainer of fighting cocks whose whole patrimony has often been lost by betting on cock fights."

Yet much Roman knowledge of raising chickens appears to have come from the Greek island of Delos, a center of a lucrative poultry production that likely provided a model for the business at home. By the height of the Roman Empire, chickens were a common sight in the bustling marketplaces. A second-century AD stone relief from Rome's port of Ostia depicts a lively market scene of a woman selling freshly slaughtered chickens that hang upside down from a wooden beam.

The Roman Empire's only surviving cookbook, the *Apicius*, features chicken in seventeen recipes, including one that calls for plucking the chicken while it is alive, a method still used in some parts of China since it is believed to make the meat tastier. The author who compiled the text around AD 500 included all parts of the chicken in his dishes, from the testicles and fatty spot where the tail feathers attach to diced brain and cooked comb. Chicken dishes even accompanied some Romans and Celts in Britain and Germany to the grave, where the bird bones mingle with human. The empire's disintegration also destroyed the Roman chicken industry, which required careful organization, well-constructed buildings, and good and safe roads. Not until the nineteenth century were conditions ripe for resurgence.

The central obstacle to mass chicken farming is reproduction. The chick embryo takes three weeks to come to term as the mother incubates the egg using her body warmth, rolling them three to five times a day to ensure normal development. The temperature around the egg must hover between 99 and 105 degrees Fahrenheit with relative humidity remaining close to 55 percent for most of that time, increasing in the last few days. Hens that spend so much time caring for their eggs have less time to lay more, so building up a large flock is time consuming.

The solution is an artificial incubator to mimic the hen and divert the bird to more egg laying. The West, however, was oddly slow to grasp the basic technology and adapt it to European and American conditions. Jefferson complained in 1812 to a friend that inventors had yet to focus on the practical application of science to incubation. But ancient Egyptians and Chinese had practiced artificial incubation as early as the fourth century BC, by heating large rooms with fires of straw and camel dung as attendants turned the eggs or by covering them with rotting manure that would provide the necessary heat. The technique appears to have initially applied primarily to goose eggs in Egypt and duck eggs in China. Over time, however, the chicken displaced other birds, as hens can lay more than waterfowl, and their eggs hatch faster.

By the medieval era, Europeans marveled at multichambered incubators in the Nile delta capable of handling thousands of eggs at a time. The exact techniques, however, were closely guarded secrets passed down within a few Coptic families over generations. The Arabic term for *incubator* translated as "chicken machine." A Medici managed to bring an Egyptian to Renaissance Florence to set up an incubator, and two French kings funded efforts in the fifteenth and sixteenth centuries to increase poultry production with the help of artificial breeding facilities. None was successful, perhaps because of the chilly and unstable European climate or the notable lack of camel dung.

The French polymath René-Antoine Ferchault de Réaumur visited Egypt, and estimated that the 386 ovens he counted could hatch an astonishing 92 million chicks a year. No Western country could match that sort of production until the twentieth century. Réaumur developed an innovative incubator heated by a woodstove governed by a sensitive thermometer. The resulting chicks delighted French king Louis XV, but the invention proved economically impractical.

New technologies in the late nineteenth century made possible efficient incubators capable of handling thousands of eggs. This development coincided with a wave of European immigrants fleeing west. In 1880, there were 100 million chickens in the United States producing 5.5 billion eggs worth $150 million. A decade later, more than 280 million chickens laid 10 billion eggs worth $275 million. New York City Jews were the single most concentrated market for chicken. The Talmud calls for Orthodox believers to enjoy the Sabbath, and that includes eating well. Beef, however, was expensive. While fish was the most popular main dish among those who fled pogroms and poverty, chicken was a close second. This preference began to drive the nascent poultry industry. Between 1880 and 1914, some 2 million Jews—one-third of the entire Jewish population of Eastern Europe—arrived in the United States, and most gravitated to New York City.

At first, most of the city's supply of live poultry was brought by boat from the Virginia coast, but by the 1870s the first trains filled with live chickens arrived from the Midwest at the old West Washington Market in Manhattan. Eggs were so valuable in the burgeoning city that many were imported from abroad. "Two hundred thousand dozen eggs have been received at this port from Europe during the past nine months," reported the *New York Times* in 1883. "The eggs came packed in straw in long cases containing 120 dozen each." As African Americans dominated poultry markets in the antebellum

South, Jews—particularly Jewish women—became the most successful poultry merchants in the nation's largest city.

Jewish housewives typically would buy a live chicken from a market and then take it to a kosher butcher for ritual slaughter, but the
enormous volume of birds often made a mockery of laws designed to
prevent suffering of the animal—such as use of a razor-sharp knife—
and honor its sacrifice. One horrified rabbi wrote in 1887 of "rivers
of mud, mire and blood" so crowded with people and birds that the
butchers "lack room to turn" and must slay birds without time to
test the knife's sharpness. In 1900, there were fifteen hundred kosher
butcher shops in New York City, and two thousand train cars of live
chickens arrived in Manhattan, a figure that quadrupled by 1920.

African Americans began to flee their own pogroms and poverty
in the rural South. Jim Crow laws and the cotton-eating boll weevil
appeared simultaneously in the 1890s, forcing tens of thousands to
leave their homes in search of a livelihood in the factory towns of
the industrial North. Humiliated as they fled, since trains in that era
denied them entry to white-only dining cars, they often had to pack
their own food, and the West African tradition of fried fowl came
with them. The emigrants called the trains "The Chicken Bone Express" for the remains of meals littering the tracks that carried them,
as well as their expertise and love of the bird, to the expanding cities.

The growing urban need for meat and eggs rippled out to farms
across Appalachia, the Midwest, and California, and rural women
were key to meeting the demand. As late as the 1930s in the United
States, women were the primary caretakers of chickens, as they still
are in developing nations today. Men busy planting cash-crop fields
or herding cattle left chickens to their mothers, wives, and daughters.
After all, the bird was a homebody. Fed on leftovers, kitchen and
garden scraps, and barnyard bugs, chickens on small farms prefer to
stick close to home.

The South lagged far behind California and the Midwest in
the poultry business at the turn of the last century, when chickens
emerged as the nation's most profitable agricultural business behind
corn and cattle. Although 90 percent of North Carolina farms had

chickens, the average flock was only twenty-two birds; the average flock in the Midwest was almost three times as large. Since poultry was considered women's work and the bird was associated with African Americans, even the promise of quick profits failed to entice many Southern white men to convert their dirt farms to chicken production. One North Carolina extension agent complained in 1917 that his exhortations were met with ridicule. "Poultry is looked upon on most farms as being an unnecessary nuisance."

Southern women were quick to see the financial possibilities, however, in part thanks to growing literacy and the spread of cheap farming periodicals. In 1909, a teenager named Mollie Tugman in the mountains of western North Carolina announced to her family that she intended to make more money from poultry than the family made growing corn in the rocky hills in one of the poorest places in the United States. Her father and older brothers laughed at her. Ignoring their guffaws, Tugman enlisted her eleven-year-old brother and two oxen and set to work building a fenced yard and chicken coop. Impressed by her determination, the family eventually gave in and lent a hand.

Tugman was inspired by articles—often written by women—in the *Progressive Farmer* that passed on tips about poultry rearing. So she knew to grow grain to feed her birds rather than give them just scraps and rougher corn, and to keep an eye out for disease. Once her flock was laying, she bought a chemical called water glass to preserve eggs, which she could then sell in the winter for much higher prices. When she married the following year, Tugman was on her way to making an income that gave her a large measure of independence unusual for a woman of her day.

Another Carolinian named H. P. McPherson wrote in 1907 that chickens provided greater profit than vegetables, butter, or milk, and relieved her of "a feeling of dependence." She concluded "there is little excuse for a woman to be without money if she has room to raise poultry." Between 1910 and 1920, gross sales of North Carolina chickens and eggs doubled, and then doubled again in the following decade. Poultry was the state's fastest-growing livestock business.

World War I transformed the chicken from barnyard adjunct to national security necessity. Herbert Hoover, head of the effort to keep U.S. troops and European civilians fed, encouraged Americans to raise backyard birds for both patriotism and profit. "Save hens, raise hens, eat eggs, eat roosters!" crowed an April 1918 edition of the *San Francisco Chronicle*. An American soldier holds his bayoneted rifle as a perch for a hefty chicken. "The more eggs the more beef and pork to Europe. The more meat on the ships the more man-killers under the Stars and Stripes. The more soldiers the more near the end of the baby-killer of Potsdam. Johnny, get your hen!" Another poster featuring a hen stated: "In Time of Peace a Profitable Recreation, In Time of War a Patriotic Duty." Egg prices doubled.

That March, the post office agreed to ship chicks from the country's 250 hatcheries by Express Mail, a service then controlled by the Department of War. This minor bureaucratic decision had immense consequences for the future of the chicken. Many farmers in that period argued that artificial incubation was immoral, since it separated chick from mother hen. There were bitter battles in state legislatures to ban the sale of chicks less than six weeks old and forbid their transport via the mail. The War Department's move suddenly opened the vast national market to hatcheries, and resistance to incubation collapsed. A decade later, there were ten thousand hatcheries, most concentrated in Northern California and western Missouri, and more than half the nation's chicks pecked through their shells in an incubator rather than a farmyard. One entrepreneur in California wrote a 1919 bestseller called *How I Made $10,000 in One Year with 4200 Hens*, the equivalent of $125,000 today.

Women no longer kept only a half-dozen chickens to sell the odd egg. Church cooperatives sprang up to market eggs and dress hens to be shipped to urban areas. The wife of a textile manufacturer in western North Carolina owned a large hatchery that could handle a hundred thousand eggs. And a woman is credited with founding an entirely new line of business devoted solely to raising chickens for meat. In Delaware, Celia Steele ordered fifty chicks from a hatchery in 1923. By mistake she received five hundred. Instead of sending them back, she housed them in a small square wooden building furnished with only a coal stove. Steele and her coastguardsman husband fed them until they were large enough to sell for meat. With the profits, she boldly ordered a thousand more of what are now called broilers. Most were shipped to Jewish markets in New York City that also were largely run by women.

In 1925, farmers on the Delmarva Peninsula between Chesapeake Bay and the Atlantic Ocean raised fifty thousand chickens. Within a decade, that figure had grown to 7 million. A U.S. Department of Agriculture study in the 1930s estimated that eight in ten chickens arriving in New York were bought by Jewish customers who quadrupled from 500,000 in 1900 to nearly 2 million by 1930. A quarter century later, the modest wooden building where Steele raised her first broilers was placed on the National Register of Historic Places. In 1928, the Republican National Committee put out an ad backing presidential candidate Hoover, titled "A Chicken for Every Pot." The phrase harked back to the sixteenth-century pledge by French king Henry IV. This time it seemed attainable. African Americans and rural white women slowly were edged out of the business, as poultry science departments sprang up at universities and Wall Street financiers took note of the profits to be made with the lowly chicken.

The Great Depression finally convinced many Southern men to heed their wives' advice and turn to poultry. Synthetic fabrics and the boll weevil killed King Cotton, the only source of income for many farmers across the region, and the bird became a lifeline for desperately poor families. What had begun as "an unimportant farm chore—throwing out a little corn and collecting a few eggs," according to a 1933 issue of the *Progressive Farmer*, was now "a scientific business and a major source of farm income." By then, ten thousand railroad cars filled with live poultry clicked north each year to the industrial cities that, despite the economic downturn, still required a steady flow of meat and eggs. The chicken was still considered less desirable and tasty than red meat, and it remained more expensive, but on the eve of World War II, it was poised to enter the center ring of the American diet.

← ←

On a sunny June day in 1951, ten thousand chicken fans filled Razor-back Stadium at the University of Arkansas in Fayetteville in the culmination of a nationwide effort to create the fowl of the future. As a band played and the crowd cheered, U.S. vice president Alben Barkley handed a California farmer named Charles Vantress a five-thousand-dollar check for his winning entry to the Chicken of To-morrow.

The award marked the rise of a vast new industry and the meta-morphosis of the backyard bird into a technological wonder akin to missiles, the transistor, and the thermonuclear weapon that had been tested for the first time six weeks earlier. The winning bird was chosen not for its exotic stature or pure breeding but for its similarity to a wax model of the perfect carcass as devised by a team of poultry scientists. The grilled chicken in your sandwich or wrap comes from a descendant of the Vantress bird.

Like the bomb, the Chicken of Tomorrow was the child of World War II. Beef and pork were rationed during the conflict to feed the troops, but chicken was good enough for civilians, so the federal government set high poultry prices to encourage farmers to produce more poultry for the home front. Unlike in World War I, attention was focused not just on eggs but on chicken meat itself, given the sudden recent growth of the broiler industry. As a result, a black market in fowl sprang up while beef and pork stocks dwindled and the conflict on both European and Pacific fronts dragged on.

With *Chicken Little* playing in theaters, President Franklin Roose-velt organized the War Food Administration to cope with shortages. The agency promptly seized all the broiler chickens on Delmarva. Made up of Delaware and parts of Maryland and Virginia, this was the national center of poultry production, where Steele's broiler business began. Soon chicken was standard fare for wounded and recuperating veterans, and in training camps across the South, black cooks introduced thousands of young Northern and Western sol-diers to the joys of fried chicken. Propaganda posters drummed into civilians the virtues of keeping backyard poultry, and defense chickens were said to be churned out at Flockheed, a jokey take on the warplane manufacturer Lockheed. But the internment of Japanese Americans, who made up a large percentage of those highly skilled workers who could determine if a young chick was a hen or a rooster, led to an unanticipated poultry crisis. "War conditions are creating an extreme shortage of competent sexers," one company noted.

By the time the war ended, Americans were eating nearly three times as much chicken as they had at the start, but the meat came from a dramatic increase in industrial-sized farms. Huge hatcheries churned out thousands of chicks, which were shipped to farmers, who raised them in vast sheds until they were ready to be sent to slaughterhouses and prepared for market. Just twenty years after Steele penned her first broiler flock into a wooden building in rural Delaware, raising chickens for meat was a matter of national security as well as a major industrial enterprise.

Just as the Manhattan Project brought together university scientists, industrial engineers, and government administrators to unlock the secrets of the atom, the Chicken of Tomorrow project drew on thousands of poultry researchers, farmers, and agriculture extension agents to fashion a new high-tech device. Unlike the bomb effort, however, the contest was anything but secret. It was the brainchild of an Iowa poultry scientist named Howard Pierce who was a senior manager at the country's largest food retailer, A&P, the Walmart of its day. In a 1945 meeting in Canada, Pierce listened to colleagues express fear that the poultry business faced calamity in the fast-approaching world where beef and pork were no longer rationed. He suggested that what industry and the consumer needed was a chicken that looked like a turkey, with a broader and thicker breast and meatier thighs and drumsticks.

Pierce convinced A&P management to sponsor a nationwide effort to realize this ambitious goal. The chain that dominated the growing trend toward supermarkets was looking for new ways of packaging food, and it began to install freezer cabinets in some larger stores in 1946 to sell frozen meat, seafood, and vegetables. Smarting from a federal conviction for conspiring to monopolize the retail food busi-ness, the company was eager to bolster its reputation. Pierce pulled together all the major American poultry organizations, two trade publications, and employees of the U.S. Department of Agriculture.

On the Chicken of Tomorrow Committee there were no women or African Americans; after three centuries of black and female con-trol, the American bird was now firmly in the hands of white male professionals. Poultry authorities created a scoring system, wax mod-els with rounded breasts, and strict rules for the contest. The goal was to draw on the expertise of small farmers as well as large commercial breeders to create the ideal broiler, with "breast meat so thick you can carve it into steaks." Given the relatively scrawny fowl of yesterday, this was a formidable objective.

To drum up interest, A&P paid for a short movie narrated by the nation's most famous newsreel reporter, Lowell Thomas, full of seri-ous men in ties and white coats examining chickens as women and blacks in the background go about more menial tasks like feeding and dressing the birds. Given the old emphasis on egg production, "rela-tively few poultrymen took steps to develop better meat-type chick-ens," Thomas explained. The committee also cosponsored a Chicken Booster Day that included a New York City banquet and screening of the 20th Century Fox movie called *Chicken Every Sunday* starring Celeste Holm and a young child actress named Natalie Wood. The real star, of course, was the reassuring meal that brought a broken family back together without breaking the bank. The bird that provided three seasons of eggs, pin money for rural women, and the occasional spe-cial dinner was recast as a serious competitor to beef and pork.

Contests in forty-two of the forty-eight states led to regional finals and two national competitions. Fertilized eggs from contestants were hatched under the same conditions, fed the same food, and given the same vaccinations. Then they were weighed, slaughtered, and dressed. Judges, recruited from universities, industry, and govern-ments, scored for "economy of production" and "dressed carcass." Vantress won both the 1948 and 1951 national contests with a breed made by crossing California Cornish males with New Hampshire females. This hardy bird with just the right mix of European and Asian genes weighed an average of more than four pounds, twice the size of a typical barnyard chicken of the day. The speed with which it became the industry norm is astonishing. As early as 1950, most commercial broilers came from this stock and those of the runners-up. A 1951 issue of the *Arkansas Agriculturalist* declared that "the day of the slick-hipped chick is over" thanks to "the leaders of the Chicken-of-Tomorrow program." Newspapers hailed the scientifically engineered birds as "these sweater girls of the barnyard."

Women and barnyards were relics of the past in the brave new world of the postwar poultry industry. Modern chickens lived indoors, ate processed feed from automated bins, consumed a host of vitamins, breathed ventilated air, and were protected from illness by vaccinations and antibiotics. The goal was to convert feed into meat as efficiently and cheaply as possible. It worked. While beef and pork prices shot up in the decade after World War II, chicken prices fell dramatically. In the new system, farmers were contractors. The company owned the hatcheries and the slaughterhouses, as well as the birds themselves, and provided the feed and medicine. The grower raised the chickens in their own coops—now long and low warehouses—until they reached their full size and were ready for shipping.

Until the early 1950s, most American flocks contained no more than two hundred chickens, about the size advocated by ancient Roman agricultural writers. In the wake of the Chicken of Tomorrow contest, farms raised tens of thousands of birds and as many as one hundred thousand. A hen that might live a dozen years on a farm now could be fattened and slaughtered in six brief weeks.

There had been nothing like this in human history. There is no record of any other major food—meat, dairy, grains, fruits, or vegetables—expanding so quickly in volume and scale. The only exception might be orange-juice concentrate, which, thanks to scientific tinkering and clever advertising, expanded rapidly in this same period. Advances in nutrition and breeding techniques made it possible to grow a bird in half the time possible in 1940, as the price per pound plummeted from sixty-five to twenty-nine cents.

What made chickens different from, say, cows? With a long, entrenched history, ranchers were slow to embrace academic genetics and corporate methods and were generally suspicious of radical change. A rising generation of poultry magnates by contrast happily drew on the extensive research by scientists on chicken genetics to create a more efficient product. Most of these new chicken magnates were not farmers but the middlemen who shuttled birds from farm to city.

John Tyson, for example, founder of the nation's largest poultry company, which now is the world's largest meat producer, began as an independent trucker. Barely scraping by in the early days of the Depression, he started hauling broilers from Arkansas to Kansas City, St. Louis, and Chicago. By providing feed and water to the birds—a novelty at the time—he could transport them over longer distances. During World War II, as demand for chicken skyrocketed, Tyson bought up hatcheries and feed factories as well as broiler houses from failed growers, pioneering the assembly-line system that is at the heart of today's modern business.

As chickens were concentrated in huge numbers, disease could sweep through and wipe out whole flocks, while feed prices could fluctuate wildly. Only the largest operations survived and thrived, and Tyson earned a reputation as a smart and hard-nosed entrepreneur apt to fly into rages. He was not sentimental about poultry. "Just keep it simple," he said. "Kill the chickens, sell 'em, and make some money." Tyson's son Don studied agricultural nutrition at the University of Arkansas before joining the company as general manager in 1952. The father-and-son team drew on the latest science in feed, genetics, and management to expand their operations throughout the 1950s. Vitamins, vaccines, and antibiotics became essential elements of success. The fact that the product was chicken was almost incidental. "We're not committed to the broiler business as such," Don later told one interviewer. "We're committed to so many dollars invested on dollars returned on that investment."

The Tysons benefited from the Chicken of Tomorrow, which supplanted the traditional breeding culture engendered by hen fever a century before. Older varieties began to vanish. Just as auto manufacturers required uniform parts, these new industrialists wanted a bird that matured quickly using as little feed and with as little variation as possible. So the new generation of scientific breeders focused on creating hybrids with a biological lock that ensured this uniformity. This approach assured a high-yielding, predictable product, but it also meant that growers could not breed their birds to produce the same uniform results. Like hybrid corn strains, farmers had to buy the next generation from the companies controlling the genetic traits. Breeding companies kept these traits under lock and key, as classified as nuclear weapons or Colonel Sanders's famous Kentucky Fried Chicken recipe.

A few livestock experts expressed alarm at the pace of change. "Modern science . . . threatens to become dogma," warned one in 1960. "The scientist might well be advised to go occasionally to the farmyard to learn rather than teach, or, what is far less excusable, to preach." The new approach to poultry, he added, was being applied "too uncritically and with too great haste." By then, more than 70 percent of the business was in the South.

The thriving poultry industry brought jobs to some of the country's poorest regions, from Arkansas's Ozarks to the hills of north Georgia. Friends in Washington ensured minimal federal oversight. Arkansas senator William Fulbright became the industry's vocal supporter in Congress. When a bill to tighten inspections was under consideration, Don Tyson wrote a brief note to Fulbright. "Bill, this would hurt the chicken business." The proposed legislation swiftly died. By 1960, 95 percent of Arkansas growers were under contract with major corporations like Tyson. While they muttered about their status as modern-day sharecroppers, public complaints could lead to cancellation of those contracts and immediate bankruptcy. Attempts by growers and poultry plant workers to organize never went very far in the Midwest or South.

For the first time in American history, chicken was cheaper than beef or pork and available neatly packaged according to cut. Picking out pin feathers, removing the guts, and chopping the feet off chickens had long been a laborious chore for housewives in cities as well as in the country. Now they did not have to buy an entire chicken, which made the bird increasingly popular for meals beyond just an elaborate Sunday dinner. The chicken proved ideal for America's postwar boom, since it could be easily packaged. "It soon became obvious that the meat-type chicken and modern-day supermarkets were meant for each other," one poultry expert observed. Meanwhile, as people became more conscious of the dangers of fat in red meat, the low-fat bird became a more appealing choice.

Tyson's $10 million in net sales in 1960 topped $60 million by the end of that decade, reflecting that shift in consumer tastes. The actual number of broiler chickens in the United States at any one time remained remarkably stable after the start of World War II. But each bird weighed twice as much and required half the feed and half the time to mature. And the number of chicken farms plummeted from more than 5 million to half a million by 1970.

Falling prices and thin profit margins, however, left the industry scrambling to come up with ever-new ways to sell their prosaic product, from frozen dinners to precooked army rations. Hungry markets opened up overseas, and by 1960 more than 100 million pounds of U.S. chicken were shipped annually to West Germany alone. Tyson, like other major poultry operations, began to spread its product as well as its plants and way of doing business to Mexico, Europe, Asia, and South America. Upstart businessmen like Frank Perdue on the Delmarva Peninsula pioneered advertising campaigns that branded chickens, even though broilers across the country are nearly identical—a marketing ploy that the beef industry has yet to copy successfully.

Perdue lay to rest any qualms about the masculine nature of his business that for so long was disdained as women's work. "It takes a tough man to make a tender chicken," he famously told television audiences in the 1970s, a century after the industry made its first halting steps. It also still takes women and minorities behind the scenes. More than half of the nation's 250,000 poultry workers are women, 50 percent are Latino, and an estimated one in five is an illegal immigrant. It is often ugly, low-paid, and dangerous work, as is well documented in newspaper articles, government reports, and books by writers who worked undercover in poultry plants. But it makes cheap chicken widely available for consumers, including those who wait for sales on boneless breasts to stock their freezers.

Fifty years after the Chicken of Tomorrow contest, chicken overtook beef as the meat of choice among Americans. The 1980s introduction of McDonald's Chicken McNuggets and other highly processed poultry—tenders, patties, hot dogs—helped push the bird over the top. Food scientists discovered that the meat, like the bird of old, was infinitely versatile, absorbing flavors more readily than pork or beef and perfectly suited for fast food. By 2001, the average American ate more than eighty pounds of chicken a year, quadruple the 1950 amount.

The figure now is close to one hundred pounds. In 2012, Tyson recorded a third of a trillion dollars in sales, and its weekly production topped 41 million chickens in sixty plants. The broiler business is booming in the United States and abroad. The vertical integration model pioneered by Tyson has spread to rapidly urbanizing South America, India, and China, and now the cattle and pork industries are rushing to copy the approach. Once ignored and despised by many in the farm sector, poultry is now an international multibillion-dollar complex that is setting the pace for the world's agribusiness.

In Fayetteville, just off the Fulbright Expressway, a historical marker stands on the campus of the University of Arkansas near the site of the Chicken of Tomorrow's final contest. The sign commemorates the "entrepreneurs who built Arkansas' poultry industry into a major force in the world economy." The marker near Razorback Stadium stands on Maple Street at the entrance of the John W. Tyson Building, an impressive modern complex of a hundred laboratories, a ten-thousand-square-foot pilot processing plant, a host of classrooms, and, as its brochure notes, "tasting booths for sensory evaluation." Funded by the federal government, poultry companies, and a state bond approved by public referendum, it is modern and clean, a $20 million concrete-and-steel monument to the success of science and industry. There's not a live chicken in sight.

11

Gallus Archipelago

Wilhelm had a queer feeling about the chicken industry, that it was sinister. On the road, he frequently passed chicken farms. Those big, rambling, wooden buildings out in the neglected fields; they were like prisons. The lights burned all night in them to cheat the poor hens into laying. Then the slaughter. Pile all the coops of the slaughtered on end, and in one week they'd go higher than Mount Everest or Mount Serenity. The blood filling the Gulf of Mexico. The chicken shit, acid, burning the earth.

—Saul Bellow, *Seize the Day*

The modern chicken has a model number rather than a name. There is the Ross 308, the Hubbard Flex, and the Cobb 500. This last is touted by its maker as "the world's most efficient broiler" with "the lowest feed conversion, best growth rate and an ability to thrive on low density, less costly nutrition." As a result, this bird offers "the competitive advantage of the lowest cost per kilogram or pound of live-weight produced for the growing customer base worldwide."

Like cars, new models are periodically introduced. The Cobb 700 made its debut in 2007 as a slightly souped-up version of the 500. It is designed for a fast-growing South American market that craves the highest yield for the lowest possible price. In 2010, as the backyard chicken craze took hold, there was the CobbSasso 150, "ideally suited to traditional, free-range, and organic farming." All new models are carefully engineered to suit their particular market.

The world looks to the United States for the latest chicken as it used to await the arrival of next year's Chevy or Oldsmobile. Three major breeding companies control the stock of more than 80 percent of broiler chickens, and two are American. The Cobb 700, for example, is the product of the Tyson-owned Cobb-Vantress. Based in Arkansas like its parent company, it has subsidiaries around the world but began in 1916 as a small New England operation that later absorbed the line of birds created by Vantress for the Chicken of Tomorrow contest.

More than three hundred U.S. breeder hatcheries produced more than 9 billion broiler chicks in 2010. The market weight of broilers has been edging up for decades, while death rates and the amount of feed required have dropped. In 1950, before the Chicken of Tomorrow took hold, a broiler required an average of seventy days to reach the average weight of 3.1 pounds, with three pounds of feed needed per pound of bird. In 2010, only forty-seven days were needed to make a 5.7-pound bird that needed less than two pounds of feed. This revolution was not solely due to breeding. Chickens, particularly those crowded together, are subject to a legion of diseases. New vaccines based on intensive research into chicken illness were the key to reducing mortality rates in this same sixty-year period by half, to 4 percent. Improved nutrition, particularly the addition of key vitamins to feed, was the third factor in this transformation of yesterday's bird into the chicken of today.

No other livestock program in history has come close to matching that steady increase in productivity and decrease in feed costs. The result is a testament to the bird's living organic clay that so impressed William Beebe a century ago. Yet the overriding goal of producing maximum meat with minimum feed has come at costs that are not apparent to consumers. During the 1990s, for example, strains of rapist roosters spread through the facilities that breed broilers. Aggressive males, either unable to understand the usual courtship dance or incapable of mating because of their huge breasts, took out their fury on hens, sometimes killing them in the process.

Leg and hip ailments brought on by inadequate bone structure hinder health in an animal designed to put on meat so fast the developing skeleton can't keep up. Some broilers can't walk to their water and feed stations, and there are signs that large numbers are in chronic pain. One study showed that broilers learned to self-administer a pain reliever by choosing food containing the drug over food that lacked it. And there is always room for a new idea. An Israeli team recently developed a featherless chicken to reduce processing costs, using a mutant strain bred in California. Images of the bald bird sparked outrage around the world. Other scientists pointed out that a chicken without feathers is an impractical option, since it could injure its partner during mating, be more vulnerable to skin diseases, and have increased sensitivity to temperature fluctuations.

Academic researchers such as Ian Duncan at Canada's University of Guelph blame broiler genetic problems on the industry's fixation with developing birds with ever-bigger breasts. The results of modern selection have maximized profit and kept consumer costs low, but at the expense of individual birds. Even industry representatives say that they may be reaching the limits to reducing feed while increasing weight productivity in chickens.

What the French call with Gallic candor the industrial chicken is the end point of the breeding frenzy that began with The Fancy. The bird that once delighted or awed or healed today raises uncomfortable questions about who we are, what we eat, and how we should care for and relate to animals. You don't have to be a vegan to wonder if it is right to put another entire species in perpetual pain in order to satisfy a craving for chicken salad and deviled eggs. As a longtime meat eater, I found myself wishing to avoid books on my shelf with disturbing and depressing titles like *Prisoned Chickens, Poisoned Eggs; Their Fate Is Our Fate;* or *Chicken: The Dangerous Transformation of America's Favorite Food.*

Finally I took a road trip that began in Washington, D.C., to visit the National Chicken Council and explore the territory of the modern industrial bird. The council has a name with a serious giggle factor, suggesting a *New Yorker* cartoon with cigar-chomping fowl gathered around a boardroom table. The hushed offices high in a glass-and-steel building a few blocks from the White House are staid and resolutely chickenless. The conference room has little stuffed cows sporting slogans urging visitors to eat more chicken, part of a successful campaign by the fast-food company Chick-fil-A to convert beef lovers.

Founded in 1954 by broiler industry magnates—all white, male, and mostly Southern—the council represents 95 percent of the U.S. companies involved in chicken meat production. This sprawling industry employs three hundred thousand workers who annually turn 9 billion birds into 37 billion pounds of chicken that consumers around the world will spend $70 billion to consume. Another two hundred thousand people on thirty thousand farms and in thousands of trucks raise and transport the product. The American broiler industry is the largest in the world and only recently took second place to Brazil as the biggest exporter. These are meat makers with economic and political muscle that surpass that of twentieth-century cattle ranchers.

With heavyweight members like Tyson Foods, Pilgrim's Pride, and Perdue Farms, which are among the biggest food producers on the planet, the council has the ear of the capital's power brokers. Senator Chris Coons of Delaware and Johnny Isakson of Georgia— respectively a Democrat and a Republican in major chicken-producing states—recently formed a Senate Chicken Caucus to match one launched in the House of Representatives two years earlier that now boasts fifty members. The goal, Coons told council members who met in a luxury hotel ballroom near the National Mall, is to educate other senators about the contributions and concerns of U.S. chicken producers, Washington-speak for lobbying on behalf of business. "As it does in the House," council president Mike Brown helpfully explained, "the Senate caucus will give a united voice to chicken producers as we navigate the many issues of importance to our industry in the months ahead." Those issues are legion, from labor disputes to changes in federal inspection procedures to ethanol production that affects the price of chicken feed.

The go-to guy on many of the thorny issues confronting the chicken business is Bill Roenigk. He arrived in 1974 at the council during the pre–Chicken McNugget era when no chicken wings were served in bars and Americans still preferred beef to the bird. He is now a consultant but still very involved in the business. Despite his neat suit, bright tie, and shiny black shoes, Roenigk can't hide a grizzled look under bushy eyebrows when he joins me in the council's conference room. He grew up on a farm outside Pittsburgh. "Get up and help," his father told him one morning when he was ten. Along with corn and squash, they raised pigs and cows, and he did chores every morning before leaving for school, a whiff of manure trailing him in the halls. "I was snobbish about the big animals," he tells me. "I didn't want to piddle with little chickens."

Roenigk wanted to see the world, and he landed a job at the Foreign Agricultural Service at the U.S. Department of Agriculture with the hope of doing so. But his wife balked at a posting in Africa, and he found his way, in classic Washington fashion, from regulator to regulated. He helped expedite the sale of chicken legs to Russia in the early 1990s, when starvation stalked that country, and now backs the controversial idea of shipping American chicken carcasses to China, where they would be cooked, repackaged, and sent back for sale to U.S. consumers ("It's a good way to get the door open to the China market").

Like any experienced lobbyist, Roenigk is too self-assured and savvy to find himself on the defensive while discussing controversial issues. This is a top-down and tight-knit industry that has repeatedly and successfully halted efforts by the government to tighten its labor and inspection practices. It has little to fear from the carping of critics. Leaning back in his chair at the conference table, he explains affably that antibiotic use in poultry poses no threat to human health, that ventilation systems ensure quality air for broilers, and that American methods of slaughter are more humane than those used in Europe.

Yet even he acknowledges that public worries about food contamination and chicken welfare are inexorably on the rise in the United States and around the world. A delegation of worried Japanese poultry-industry officials recently visited to discuss how to cope with an increasingly outspoken animal-rights community there. Just a week earlier, a salmonella outbreak at a California plant made national headlines and sent jitters through the industry. Not long after my visit, the Federal Drug Administration banned two of three arsenic compounds used in chicken feed to improve feed digestion and then limited antibiotic use in livestock that is designed to boost productivity rather than just maintain health.

Despite political influence, the chicken business now finds itself under an increasingly relentless spotlight. It's not the quiet and behind-the-scenes position of the past, and the industry is not as adaptable as the chicken. "We must become much more transparent and show people what is going on," Roenigk says, and I believe that he is sincere. The National Chicken Council, however, cannot find a company willing to let me visit any of their dozens of plants on the Eastern Shore, which lies just across the Chesapeake Bay from Washington, though I made my request weeks in advance.

The next morning, a bright fall day, I cross the Bay Bridge, which begins at Annapolis and leads to Delmarva. One out of every fifteen American chickens lives and dies on this 170-mile-long strip of land that bulges in Delaware, tapers into Maryland, and then tails off in Virginia, coming to a point at Cape Charles, where the Atlantic Ocean meets the bay.

Though surrounded by major urban centers—to the north are Wilmington and Philadelphia, to the west are Washington and Baltimore, and to the south are Norfolk and Virginia Beach—the region of watermen and farmers remains stubbornly rural, poor, and conservative. For three-quarters of a century, the backbone of its struggling economy has been poultry. Hundreds of long broiler sheds and a dozen massive slaughterhouses owned by five companies dot the flat and swampy countryside. Every week, they process 12 million birds in a region that has barely half a million residents.

My first stop is on the drab outskirts of Dover. There, inside the

Delaware Agricultural Museum and Village, a cutout of Celia Steele stands behind chicken wire in front of a rough wooden building no bigger than a one-car garage. She is not quite middle-aged, and has full lips, a fleshy nose, and a Victorian-era blouse cinched at the sleeves. Steele's right hand reaches for something, her eyes dark and intelligent. In front of the life-sized cardboard black-and-white cutout is a little plastic sign declaring her a Sussex County poultry pioneer. It is the most modest of memorials.

Yet what began in the housewife's backyard in the little coastal town of Ocean View, fifty miles to the southeast, in fact dramatically altered the fate of two species. Steele's broiler business laid the foundation for a global industry, augured a radical change in humanity's diet, and opened a Pandora's box of vexing ethical, labor, environmental, and health issues. Chickens now produce 100 million tons of meat each year, twice what was produced only two decades ago. Egg production in the same period also has doubled. Seven billion hens lay more than a trillion eggs a year. The bird now is a critical staple of our modern industrial society. This sudden and dramatic proliferation coincides with the global migration to cities, which now hold more people than rural areas for the first time in our species' history. Steele's success, after all, hinged on the nearby market of New York City, which just two years after she began her broiler business surpassed London as the world's largest city.

The fowl is now so central to our urban lives that price hikes and supply problems can quickly transform into ominous political threats. Just as members of the British Parliament worried in the 1840s that the potato blight could spark revolution, so do politicians in developing nations today fear the wrath of a populace denied a chicken in every pot. Saudi Arabian leaders now grant generous government subsidies for imported grain to feed Saudi chickens in a move designed to keep prices low and a potentially restless population happy.

Like Londoners in the nineteenth century and New Yorkers in the twentieth, Africans, Asians, and South Americans crowding into twenty-first-century megacities like Lagos, Manila, and São Paolo are quickly developing an insatiable appetite for the bird, on a scale unimaginable in Steele's day. To feed this demand, companies are copying the Tyson approach of controlling every aspect of the supply chain. This interlocking system of large companies that breed, hatch, raise, slaughter, and market chicken involves far more than just shipments of processed meat. Chicken feed is now a misnomer, since it is a multibillion-dollar international business, as is chicken medicine and the equipment needed to house, water, feed, and process the bird.

Driving down Delmarva's highways and back roads, I don't see a single chicken. The situation would have been unthinkable in Steele's time, despite the fact that only a fraction of the fowl existed then compared with today. As chickens have become more numerous, they paradoxically have become less visible. Vast new metropolises of fowl now mirror our growing megacities. These shadow cities across Asia, Africa, Australia, the Americas, and Europe are typically set outside our urban world in the increasingly empty countryside like that of Delmarva.

Just like human municipalities, the complexes require electricity, water, food, and sewage systems, as well as road, rail, ship, and air connections to maintain and move millions of inhabitants. Many American plants, such as those in Delmarva, slaughter and process a million broilers a week. The expanded Fakieh Poultry Farms in Saudi Arabia will soon be producing 1 million broiler chickens and 3 million eggs *a day*. Privately owned, these poultry polities are largely off-limits to outsiders, and regulations vary widely from country to country. It is now possible to live an urban life eating chicken meat and eggs daily and never see an egg laid, witness a single slaughter, or even glimpse a live bird.

The abrupt segregation of the two species is almost as remarkable as the chicken's population explosion. "The attitude of a gentleman towards animals is this," said the Chinese sage Mencius some twenty-five hundred years ago. "Once having seen them alive, he cannot bear to see them die, and once having heard their cry, he cannot bear to eat their flesh. This is why the gentleman keeps his distance from the kitchen." Such distance was long a luxury available only to the rich. In his *Remembrance of Things Past*, Marcel Proust recalls the cook's ability to make the flesh of a roast chicken "so unctuous and so tender" with a skin "gold-embroidered like a chasuble." Yet one day he spies her savagely killing a bird in the courtyard while crying, "Filthy creature! Filthy creature!" He would have insisted on her immediate dismissal, except that he could not bear to lose the sumptuous meal she made with the slaughtered animal. "And, as it happened, everyone else had to make the same cowardly reckoning."

Long after World War II, the slaughtered and plucked bird typically came into American kitchens with head, feet, and guts intact—what is still called a New York dressed chicken. That changed in the late 1960s, when Holly Farms introduced packaged pieces. Even those recognizable parts—breast, wings, and thighs—have since proliferated into hundreds of products that bear no obvious resemblance to chicken anatomy, like the Chicken McNugget.

While we eat far more chicken than we did in Steele's time, we know it far less. Traits long admired by humans, such as courage and devotion to family, no longer are modeled by the bird. When we get our hackles up or feel henpecked, there is no visceral connection to the cockpit or barnyard. While pigs and cows are turned into pork and beef, *chicken* today is more likely to refer to meat than an animal. The bird has vanished from our sight even as it proliferated. When it lands in our shopping basket or on our plate, we can only trust that it was humanely handled, safely processed, and carefully inspected.

Driving south from Dover into Sussex County, I rendezvous with Bill Brown, an extension agent for the University of Delaware, at the site of the first Chicken of Tomorrow contest, in 1948, near the little town of Georgetown. He's agreed with only a couple of hours' notice to give me a brief tour of a miniature broiler house at this rural research station set amid flat fields. Brown, a stocky man with a goatee, rummages through his trunk and hands me a white hazmat-style suit. Disease rather than predators is the primary threat to today's meat birds, since they are concentrated in large numbers and harvested before their immune systems can fully develop. What I carry on my shoes could kill them all. Once we are dressed, he leads me through a metal door and into an open warehouse. "It's a small flock," he says apologetically. "Maybe twenty-four hundred birds." He's not being facetious. At home, Brown and his wife raise two hundred thousand chickens in a half-dozen houses under contract with Perdue. Even that is considered a modest number today.

A warm and humid wave of air swiftly replaces the October chill amid the high-pitched peeping of day-old chicks. Bright overhead lights reveal long pipes punctuated by bright red feed and water stations crowded with the fuzzy yellow chicks. The space seems roomy now, Brown says, but it will quickly fill up as the chicks quadruple their weight in the first week. In less than two months, they will head to a processing plant, still three months shy of sexual maturity. As Brown steps gently forward, the animals seem unafraid. He scoops one up and holds it in his palm as he explains that these birds are part of an experiment to gather data on the ammonia fumes that already make my eyes smart. The noxious gas is a byproduct of chick urine, and poultry scientists want to find better ways to neutralize its effects by altering the feed or by precipitating it out of the air. "You get used to it," he says as I blink furiously.

We step outside and I take a grateful lungful of clean air in the sudden quiet. Brown tells me that the toughest problem is coping with the chicks' waste after they are sent to slaughter and the sheds are cleaned. The 600 million chickens raised on Delmarva produce more waste than Los Angeles. The manure is rich in phosphorus and potassium and more than twice as effective as commercial fertilizer, well suited for nutrient-poor fields. The manure's high level of phosphorus was thought by some researchers to bind soundly with alum, lime, and other substances, preventing damaging runoff into streams and rivers. But now soil scientists say that the water table can absorb and move phosphorus in ways that are still poorly understood. And with plant expansions, there is too much manure and too few fields on the peninsula to cope with the mounting waste. Watermen fear it is killing their crabs and fish. Moving it is expensive; Perdue experimented with turning manure into pellets that could then be shipped to the Midwest but abandoned the effort as it was too costly, while attempts to turn the waste into gas have yet to bear fruit. The confusing county and state laws make a solution to the growing problem both politically unlikely and technically elusive.

Continuing south, I pass through Maryland and into the narrowing strip that is Virginia's Eastern Shore, stopping for a lunch of soft-shell crab just outside Temperanceville, a village of less than four hundred people where no whiskey has been openly sold since 1824. A few miles on, on the right side of the highway, a low-slung building as long as a football field fronts a wide manicured lawn. This is part of a Tyson plant employing more than a thousand workers to slaughter and package a million chickens each week. Environmentalists singled out this plant as the most polluting poultry facility during the 1990s amid a series of devastating fish kills in Chesapeake Bay. Without permission to visit, I drive on for another forty miles, almost to the tip of Delmarva, pass a massive Perdue facility, exit the highway, pass rusting trailers with yards full of children's toys, and rumble down the grassy lane of a 1920s colonial-revival house that is the headquarters of United Poultry Concerns.

Though invited, I am apprehensive. When I first contacted Karen Davis, the founder and sole full-time employee of the animal-rights organization, she shot back a curt email response, calling a magazine article I did on the bird "despicable." What I needed, she wrote, was "a whole different perspective, spirit, and attitude toward chickens." But she would make time to meet me. I swallow my irritation and knock on her door. Instead of encountering an angry crusader, however, I meet a thin, smiling woman with a thick mop of black hair. On her matching black Windbreaker is a button that says, "Stick up for Chickens."

The lecture I expect doesn't come. Instead she suggests that I meet her birds, as a parent might proudly offer to show off her young children. I follow her through the living room and kitchen—every horizontal and vertical space taken up with chicken knickknacks, prints, and posters—onto a porch, and into the backyard. There are tall cages amid a large fenced yard bordered by shade trees. She introduces me to Ms. Hendy, Pace, Petunia, Taffy, Buffy, and Biscuit, and proffers the story of each—this one rescued from the Perdue plant, another from a lab in Norfolk, yet another from a raid on a Mississippi cockfighting ring—then brandishes a rake to protect me from the hot-tempered rooster Bisquick. Some are left anonymously on her doorstep. Since 1998, when she moved here from the Washington suburbs, her modest two-acre sanctuary has been an island for misfit poultry.

After the numbing uniformity inside the Delaware broiler shed, the individuality of each of Davis's birds is startling and unnerving. Beside the spry and lean gamecock, the rescued factory chicken with its pumped-up breast and spindly legs looks grotesque. Commercial laying hens typically have the tips of their beaks seared off to prevent them from injuring their neighbors, and I spot one pecking pathetically in the grass, apparently unable to spear a worm with its defective bill. An aging broiler saved from the slaughterhouse sits in the corner of a cage, too heavy and lame to walk. "The industry has created chickens that have chronic pain in order to get birds that grow at the far outer limits of what is biologically possible," writes Temple Grandin, a Colorado animal scientist who has pioneered animal-welfare systems. Davis's backyard is a stark lesson in the grim realities of the modern chicken.

Davis knows from long experience as an activist that her folksy tour is far more effective than any lecture. The litany of horrors and illusions is well documented in peer-reviewed scientific journals, animal-rights periodicals, and books. Debeaking, even if it is not botched, is a painful procedure typically accomplished with a heated blade that deprives a bird of its primary sense organ. A significant percentage of birds that are not killed by the knife end up scalded to death in hot vats. Most U.S. free-range and organic chickens, like their industrial cousins, never see the sun, eat a worm, choose a mate, or raise a chick. And, I'm surprised to find, poultry grown for food is exempted from any and all U.S. government rules regulating animal welfare. "Is it any wonder that many people regard chickens as some sort of weird chimerical concoction comprising a vegetable and a machine?" asks Davis. Seeing the damaged survivors of the system brings home the extent of the problem.

We mount the porch steps, and Davis points out an area reserved for aging birds, a sort of poultry rest home. In her living room full of chicken tchotchkes, she tells me that she didn't grow up around the birds, though she still broods over a dark memory of a neighbor killing a fowl during her childhood in Altoona, Pennsylvania, just east of the farm where Roenigk grew up. Drawn to the animal-rights movement in the 1980s, Davis quickly became enchanted with chickens. "Guinea fowl aren't going to sit with you or perch on your shoulder, but chickens can be very affectionate," she tells me. "They are cheerful, friendly, and like to be with you. They look you in the eye, and one of their charms is that they also live an autonomous social life—they aren't waiting for you to release them from boredom. And," she adds with a laugh, "they have the most balletic way of tripping through the grass."

Though her poetic description is sentimental—Bisquick clearly would like a piece of me—it is as succinct an explanation for the chicken's close relationship with humans as any I've heard. She owns a sweatshirt that says, "I dream of a society where a chicken can cross the road without its motives being questioned."

Davis is an unapologetic take-no-prisoners vegan, who believes that no human needs to eat an animal or even an egg. "We can get all the protein we need from plants. We can turn them into chickenlike flavors and textures. We have all this ingenuity, and we pride ourselves on what we can do. There is no need to kill chickens." Nearing seventy, though looking much younger, she pushes legislation, keeps up on the scientific literature, posts outraged blogs, and protests every year at the Delmarva Chicken Festival and at the Brooklyn *kapparot* rituals on Yom Kippur eve.

She is realistic enough to acknowledge that her efforts will fail. "Chickens are doomed," she says without hesitation. "It's the doom of proliferation, not extinction. I think it's a worse doom than extinction. I think chickens are in hell and they are not going to get out. They already are in hell and there are just going to be more of them. As long as people want billions of eggs and millions of pounds of flesh, how can all these animal products be delivered to the millions? There will be crowding and cruelty—it is just built into the situation. You can't get away from it." She pauses again. "And we are ingesting their misery."

It is a haunting refrain that predates the Australian philosopher Peter Singer's 1975 *Animal Liberation,* which helped set the modern animal-rights movement in motion. Her words echo back to the ancient mathematician and mystic Pythagoras, who is said to have been an ethical vegetarian as the first chickens arrived on Greek shores. Devout Jains, Hindus, and Buddhists for millennia have avoided meat, given their belief that animals share soul material with humans. "No, for the sake of a little flesh we deprive them of sun, of light, of the duration of life to which they are entitled by birth and being," the first-century AD Greek essayist Plutarch wrote about farm animals. "The eating of flesh is not only physically against nature, but it also makes us spiritually coarse . . . it is perfectly evident that it is not for nourishment or need or necessity."

More recently, the writer J. M. Coetzee warned that "we are surrounded by an enterprise of degradation, cruelty and killing which rivals anything that the Third Reich was capable of, indeed dwarfs it, in that ours is an enterprise without end, self-regenerating, bringing rabbits, rats, poultry, livestock ceaselessly into the world for the purpose of killing them." Miserable and tortured animals will not in the long run make for a happier humanity and a better world. Most of us would be appalled, as Plutarch was, at the ancient practice of thrusting red-hot iron pikes into the throats of live pigs in the belief that it would make the flesh tenderer. What goes on in the chicken cities of our own century, though still largely shielded from public scrutiny, may someday be seen by a majority of humanity in a similar light.

Davis's stance is by her own admission impractical, her actions ineffectual, and her views wildly anthropomorphic. Her tenacity as a Don Quixote of poultry, tilting at Tyson and Perdue in their very backyard while caring for the fowl flotsam that comes her way, is undeniable, and strikes me as worthy of attention and respect. As I drive away in the autumn dusk, uncomfortably pondering my own cowardly reckoning as a chicken eater, she's heading for the backyard to shoo her flock into their cages for the night.

↙ ↙

American-style factory farms are spreading inexorably around the globe, but there are a few pockets of stubborn resistance. Shoppers in some countries are still willing to pass up less expensive imported birds in favor of the traditional varieties that they find much tastier. And the chicken still performs functions that don't require high-tech animal engineering. When a Western aid organization introduced the Rhode Island Red to rural Mali in sub-Saharan Africa, it was literally a flop. Villagers seeking a prophetic sign watch to see if a dying chicken falls to the right or left. The new birds proved useless for divination, since they fell forward on their massive breasts.

In the bastion of Bresse, in eastern France, chickens exert a nearly mythic hold that the industrial chicken has yet to break. The Latin word for rooster and France is the same—*gallus*. Though linguists say this is a coincidence, ancient Celtic tribes in the area held the cock sacred. A rooster often accompanied their god Cissonius. The bird's high status among medieval Christians in this Catholic country and its fierce fighting abilities eventually made it the national symbol. With the exception of a brief period in Napoleonic times—the emperor was not a rooster fan—eagles were left for Hapsburgs and Germans. According to one account, there was "a fierce zoological debate" within Napoleon's Council of State that ended with the choice of the rooster as the government's emblem. "It's a barnyard creature!" scoffed the emperor, who nullified their decision in favor of the Roman eagle.

During the 1789 revolution, citizens carried flags emblazoned with the cock, and in 1830 it replaced the fleur-de-lis as the national emblem. France's official seal is Liberty sitting by a ship's tiller decorated with a proud rooster. The bird appears on coins and on war memorials, surmounts the gate of the presidential Élysée Palace in Paris, and even sits atop the golden pen set on the president's desk.

One of the perks of that high office is a gift of four of the country's finest poultry for the ritual Christmas meal. The chickens come from Bresse, an old French province lodged between the city of Lyon to the southwest and the Swiss border to the east. The region has long been famed for its food. "We wired to the Picassos that we were delayed," writes Gertrude Stein's partner, Alice B. Toklas, in *The Alice B. Toklas Cookbook*. The couple was traveling south from Paris to the Mediterranean for a summer holiday with the painter and his wife in the 1920s when they encountered the countryside and food in the foothills that led up to the Swiss Alps and stopped in the bustling market town of Bourg-en-Bresse, "renowned for its chickens of the large and thick breasts and short legs."

That chicken is mentioned in Bourg-en-Bresse archives as early as 1591, when citizens bestowed two dozen on the local feudal lord for saving the city from invaders. When Henry IV—famous for his pledge to put a chicken in the Sunday pot of every peasant—showed up a decade later, he was delighted by the quality of the poultry. The celebrated gastronomical author Anthem Brillat-Savarin praised the Bresse chicken in 1825 as "the queen of poultry and the poultry of kings." In 1862, a local count organized a contest to select the best-tasting chicken from among black, gray, and white varieties. The winner was a well-proportioned white bird with a bright-red comb and dark-blue legs, its colors the colors of France. Thanks to railroads, the Bresse chicken soon was a favorite dish among the elite from Paris to St. Petersburg. Along with regular birds, the farmers specialized in the ancient and difficult art of castrating young roosters to make tender capons and sterilizing young hens called poulardes.

The fowl drew the attention of Darwin's collaborator William Tegetmeier, who at first wasn't impressed by what looked like an ordinary European bird, inferior in size to the big new Asian varieties. But a French colleague explained to him in 1867 that the secret to their superb taste was "the skill and the habits handed down from one generation to another among the farmers of La Bresse." The special property of the Bresse soil; a careful regimen of local corn, wheat, and whey; and an unusual method of slaughter were the key. Flocks were kept small to avoid disease, the birds were placed in cages for the last couple of weeks of their lives to fatten them up, and a knife driven quickly into their palate bled them efficiently.

In the era of hen fever, Bresse farmers crossed Asian chickens like the Cochin with their smaller birds, and the result was a larger but coarser-tasting chicken. Rather than sacrifice taste for quantity—a choice capon in Tegetmeier's day might sell for the equivalent of a hundred dollars today—they passed the next century and a half quietly sticking to their old ways and traditional breed. Every December, four contests, grandly called the Les Glorieuses de Bresse, are held across the region to choose the finest chickens, which arrive at the Élysée Palace by Christmas Eve with the kind of pomp that Americans reserve for the lighting of the tree at New York's Rockefeller Center.

Today, the Bresse chicken is in the august company of the sparkling wine called Champagne and the rich cheese called Roquefort. Granted an *appellation d'origine contrôlée*, the bird is now licensed to ensure that it was made in the region under exacting and traditional standards. This required an act by the French National Assembly in 1957. It is the first and only animal to receive such a designation, apart from the salt-marsh lamb of the Somme that in 2006 was also granted a comparable license. When I sit down at the kitchen table of Pascal Chanel, a poultry farmer who lives just outside Bourg-en-Bresse, I'm astonished to hear him describe methods virtually identical to those laid out by Tegetmeier's French informant at a time when Americans were still recovering from the trauma of the Civil War.

Chanel has won the Glorieuses three times, and he has the blue vase traditionally sent in thanks by the president—then Jacques Chirac—prominently displayed among other trophies and awards in a cabinet by the back door that leads to the farmyard. Clean shaven, with thinning hair and a Gallic nose, Chanel explains that his flock must come from the only Bresse chicken hatchery, and that no flock can exceed five hundred in number. After their first month, each chicken must have at least thirty square feet of open field to wander for four months. They can never be fed genetically modified grain and subsist on a carefully calculated diet of local corn, wheat, and skimmed milk. Protein is limited so that the birds will find their own amid the grassy fields on his thirty-acre farm, where he also grows their feed. "I prefer to see my poultry eat insects and worms than antibiotics," he says.

Drugs can be administered to sick birds, but this requires a written report by a veterinarian and subsequent government approval.

We leave the kitchen, warmed by a roaring fireplace, and step into a cold November drizzle. At first, I'm not sure what I'm seeing. Then I realize that the green field ahead is spotted not with faraway sheep but with nearby chickens. In all my travels, I've never seen a large flock feeding in an open field. "The main problem is predators," he says, as we pass the feathers and bones of a recent kill. One in five succumbs to foxes and hawks. But the most dangerous predators are human. Since a Bresse chicken can cost $30 per pound, and a capon might fetch a total of $275, theft is an ever-present threat.

In the final couple of weeks, the birds are herded into cages for fattening. "If they are left outside the meat is too tough," Chanel explains as I follow him to a small wooden shed. Eighty birds are in wooden cages called *épinettes*. They are like battery cages, except their wooden frames and slats give them a more rustic air. There is no burning smell of ammonia in the well-ventilated shed. Each cage holds about four chickens—there's room for each bird to stand and move around—and they poke their heads out occasionally to peck at the mash and cast a glance our way. "It's the happy time," he says through my translator, a Lyon woman in heels. I look at her, baffled. She smiles, explaining as if to a child that the birds get to rest, relax, and eat in their final ten days to two weeks. She describes the *épinettes* as a sort of spa, with dim lighting and unlimited food. One local website asserts in a charming translation that the chickens in *épinettes* "love the attention attended them with their regular meals of whey-rich porridge."

In the next wooden shed are several dozen capons. Chanel motions for me to be silent. I look through the door. They are huge, nearly twice the size of the others, and lack combs, though they also lack the misshapen breasts and legs of the American broilers I've seen. While outside, each must have sixty square feet of meadow. In the *épinette*, each has its own cage. When Chanel enters, there are some quiet clucks, but as soon as I duck my head into the room, they panic and flap their wings as one, and I quickly retreat. Without their gonads, the farmer says, they don't crow and lack the natural fierceness of roosters. Veterinarians now perform castrations, and Chanel estimates that only one in a hundred doesn't survive the operation. These nine-month-old birds are the most prized for a French Christmas dinner. After slaughter they are hand-plucked, washed, and laboriously sewn into linen corsets. This spreads the fat around the meat and limits air exposure to ensure preservation. This was essential in the days when it took two weeks to reach the Paris market, and it remains the common way to present the bird.

As we trudge back across the muddy driveway, I overhear my translator making what seem to be arrangements to come by in December to purchase a capon. "It is much cheaper to get it from him than to buy it in Lyon," she tells me. "I'll bring my kids and let them see the farm." Unlike an American grower, Chanel is a truly independent operator who can choose his customers but who also assumes the full financial risk. He says that he makes a decent living raising four thousand birds annually, so long as he doesn't count the hours. Summers are hard, since the chickens don't go to bed until dark comes around after 10 p.m. Along with caring for the birds, he also raises all their feed, gesturing at the bulging corncribs. The *appellation* requires careful record keeping and frequent government inspections. The scattered sheds required don't lend themselves to automation, and he is worried that the younger generation is losing interest in the hard work required to maintain the flocks. "It's like raising children," he says with a rueful smile. I ask him about the bus sitting beside the barn. It is part of his supplemental income. "I've been a school bus driver for twenty-five years," he explains.

About 250 farms raise 1.2 million Bresse chicken a year, only a little more than the number processed in a single week by the Tyson plant in Temperanceville. Just one in three hundred French chickens is a Bresse, and each is given a distinctive label and a metal ring around one leg. Given the high cost to produce the birds, the Bresse is reserved primarily for upscale restaurants and butchers. A 2006 avian influenza scare forced anguished farmers to lock up their birds, and the fate of the French institution was for a time in question. "The problem is to find enough producers," says Georges Blanc, chair of the Bresse poultry trade committee. Blanc—gastronomy magazines tend to attach "legendary" to his name—is a famous chef with a three-star restaurant and upscale spa in the village of Vonnas outside Bourg-en-Bresse.

Though dressed simply in his chef whites, he gives off a feudal air sitting behind a massive desk in his multimillion-dollar half-timbered complex. Across the square is a store where you can buy Georges Blanc wine, snail pâté, pheasant terrine, and tripe packed in Calvados. There is an imitation traditional dovecote in the park nearby. Our chat is more audience than interview. Blanc describes his successful effort to fight off an attempt by farmers in the 1970s to lower standards so they could sell more birds. "It was tradition versus change, and we won," he adds, a three-star culinary general describing a hard-fought battle. He makes an incredulous face when I ask him if he ever uses fowl from factory farms. "I *never* cook an industrial chicken!" Only the finest-quality products go into his dishes, Blanc adds with a touch of Gallic indignation.

Two or three times a week, he partakes of the Bresse chicken. Despite his concerns about production, Blanc is confident that the bird will continue to hold top rank among poultry aficionados who travel from as far away as Japan to sample his famous cuisine. The regional government provides start-up funds to young farmers who want to maintain the tradition. His business, he adds, is to cater to his specialized clientele, not to worry about the future of the industrial chicken.

Blanc glances at his watch, and I excuse myself. He does not ask me to stay for dinner, nor can I afford the $122 price tag for the Bresse chicken dish listed on his restaurant menu, although, granted, that includes foie gras and Champagne. Instead I return to Bourg-en-Bresse for a leisurely meal at the restaurant Le Français in the heart of town. The old-fashioned brasserie with its belle epoque decor has a livelier and friendlier feel than the stuffy Blanc dining room. Soon after I arrive, there is not an empty seat on this chilly and damp Tuesday evening in November.

What arrives on the plate is a bit of a shock—a bare roasted leg, its nub still a purplish hue, covered in a white sauce, stark and visceral, unencumbered by vegetables or other distractions. The first taste reminds me vaguely of turkey, but moister and denser. Unlike regular chicken, the skin is buttery, and even the tendons are tasty. Skeptical of the hype, I'm taken by surprise. This is chicken that, thankfully, really does not taste like chicken.

Bresse poultry will never feed the world, or even many individuals beyond Blanc, his wealthy clients, and a small number of people on a splurge. But other French labels requiring lots of outdoor space and time, careful feeding, and local control have captured a quarter of the national market. One U.S. businessman, frustrated by the industrial food system in his own country, is attempting to import that concept as part of an ambitious dream of transforming the way Americans buy and eat chicken. When I get back from Bresse, I look up Ron Joyce at Joyce Farms outside Winston-Salem, North Carolina.

He is an unlikely revolutionary. Joyce was raised on a thirty-acre farm in western North Carolina, and his father worked for Holly Farms—an early innovator in packaging chicken parts for supermarkets—before starting his own business shipping chickens to fast-food businesses like Bojangles'. Now fifty-seven, Ron took over in 1981 when his father passed away. Soon Tyson muscled into the market, buying up Holly Farms, and began battling with another titan, ConAgra Foods, for control of the growing fast-food market. Joyce sold to ConAgra and began thinking about a specialty business.

Then, a decade ago, he took a trip to Paris. "It made me angry," he says in his Carolina accent as trucks on I-40 boom past his office window. "I tasted things that I didn't know existed. The butcher shops were stacked with an astounding variety of poultry that you could buy fresh every day. I came back to the richest country in the world and we were all eating one chicken—a Cornish Cross that grows faster and faster through genetic selection and has no flavor."

Unlike many slow-food and animal activists, Joyce blames not corporate greed but consumer apathy for the sad state of poultry in the United States. "People vote with their pocketbook. In Europe, they will pay a lot more for their food. Here, you put boneless chicken breast on sale for a dollar ninety-nine and it triples your sales." Whether chicken or carrots, it all comes down to price for the average American, and industry races to oblige. Profit margins for the poultry business, compared to pharmaceutical or other manufacturers, are thin and subject to larger forces like the weather and price of corn. So companies shave costs and prices to maintain an edge over competitors. "I want the consumer to take responsibility," he says. When his kids wanted to stop recently for a thirty-nine-cent taco, Joyce shuddered. "I told them, I don't *want* a thirty-nine-cent taco. I mean, how low can it go?"

Overuse of antibiotics, adding arsenic compounds to feed to improve digestion, soaking slaughtered birds in dirty chlorinated water, and then adding chemicals to kill dangerous bacteria are the result of this relentless push to lower costs. So are birds with massive breasts and weak legs that must be euthanized because they can't stand up to get to feed and water bins. "You can't put an American chicken outside," Joyce adds, struggling to contain his outrage. "Its immune system is not developed enough in six weeks, when it is shipped to the processing plant." Such practices, particularly the chlorinated water procedure, explain why the European Union bans imports of all U.S. poultry. The National Chicken Council is trying desperately to overturn this embargo to gain access to large and profitable European markets like France.

As an alternative to this race to the bottom, and to find a niche unoccupied by the big corporations, Joyce set out to copy the French approach. He began by selling a free-range and organic chicken, but quickly realized that the genetics of American birds are all the same, no matter how you feed and care for them. The meat was still tasteless. Marinating was the only way to put flavor back in the bird. Then he discovered Label Rouge. It is Bresse-lite, with a price tag that many consumers could afford. I had seen it in the display cases of Bourg-en-Bresse shops for about half the price of its more famous cousin. "This chicken feeds outdoors, but with less room, and is slaughtered after ten or twelve weeks rather than sixteen," one butcher with a shop across the street from the town's live poultry market had explained.

First, Joyce sought out aging Carolina farmers who had raised heritage birds in the days before the Chicken of Tomorrow dominated. Then he imported eggs from France, though his French suppliers were skeptical. "They laughed and said Americans would never pay for it." Joyce imitated the Label Rouge approach to care, feeding, and hand processing. Enter the Poulet Rouge de Fermier du Piedmont, the red farm chicken of central North Carolina, though it does sound better in French. Joyce cultivates chefs along the East Coast, many of whom had no idea that the chickens they used were genetically identical to those sold to McDonald's. How much value can you add with sauce? By educating these influential players in the American food system, Joyce hopes to create a demand for a bird that both tastes better and is more humanely grown.

Imported eggs, smaller flocks, more grain, and nearly twice the growing time means a price at least twice that of a Tyson chicken. Joyce is betting that improved taste will be the main driver in bringing customers to his door. The animal welfare aspect to his system, however, is what impresses Leah Garces, who directs the U.S. office of Compassion in World Farming. She says that improving the lives of American chickens hinges on better living conditions and slower growth. Doubling their life span of forty-five days is key to ensuring adequate bone growth before they begin to put on weight. And with natural light, more space, and greater variety—straw bales, planks, and ladders give them something to do in their coops—the Joyce chickens are scored among the best treated in the industry. They are not a heritage variety, since they come from one of the three companies that controls 80 percent of all chicken breeding stock, but they are an older traditional breed called the red naked neck that is quite different from the ill-proportioned, fat-breasted, and thin-legged chickens in the United States.

Joyce introduces me to his son, who works in the next office, and the three of us go on a tour of the plant. Their casual and frank manner is a stark contrast to Big Poultry. Not only do they let me see what I want but they answer every question without hesitation. I'm welcome to take pictures and film. Our first stop is the hatchery across the parking lot, where two men are moving eggs out of an incubator. With a capacity of fifteen thousand eggs, it is tiny by industry standards. Then we walk into the slaughtering area. That job takes place in the morning, and two employees in bright-blue smocks are mopping up the gleaming stainless steel equipment as we walk through. The birds live within a one-and-a-half-hour radius, so travel time is minimal, an important factor in welfare and meat quality. The trucks back up to the loading dock and employees take the live birds, hang them upside down, stun them electrically, and then slit their throats. A hot-water bath, though not hot enough to remove flavor, loosens the feathers. The remains are then eviscerated by hand.

If a chicken doesn't go through rigor mortis, it will be tough on the plate, and U.S. plants plunge the carcass into cold chlorinated water to cool it down. The birds can absorb the water, which adds weight—and therefore profit—but they also can absorb unwelcome bacteria in the bath. Often a chemical spray is added to kill the bugs. Joyce, as do many Europeans, uses fans to air cool the birds instead. It takes a little longer, but it is a cleaner process. In the next hall, women are deboning birds by hand. The meat is packaged, boxed, and shipped to customers. Though there is modern machinery, a good deal of hand skill goes into the process. The pace is slower and the quantities produced are minuscule compared to the massive poultry processing plants in the area. Five thousand Rouge chickens leave his plant a week. Just a couple of hours away, one of the world's largest poultry plants, owned by Mountaire Farms, produces millions of chickens per month.

Yet Joyce is creating an alternative, however small and elite, in a country with shockingly few options between the industrial chicken and vegetarianism. He's branched out to include other birds such as the guinea fowl and turkey, working with breeders to improve taste while keeping welfare in mind. A few weeks after my visit to the Winston-Salem plant, I'm surprised to see ring-tailed pheasant on a restaurant menu in my own town. This is the bird imported from China in the 1880s for the benefit of Western hunters. Yes, the waitress confirms after checking with the kitchen, it comes from Joyce Farms. Confident that the bird led a better life than any industrial chicken, and grateful to have the alternative, I order it. Tender and dense but not too gamey, it is delicious.

12.

The Intuitive Physicist

I looked at the Chicken endlessly, and I wondered. What lay
behind the veil of animal secrecy?

—William Grimes, *My Fine Feathered Friend*

Hens might be skittish and roosters horny, but fowl in the ancient and medieval worlds were rarely considered unintelligent. Their role in royal menageries, religious rituals, and healing potions made them worthy of respect, emulation, and even awe. In 1847, as The Fancy took flight in Britain and America, a New York magazine published a ridiculous riddle that, however, did not ridicule the animal. "Why does a chicken cross the street?" The answer was, of course, "Because it wants to get on the other side!" In L. Frank Baum's 1907 *Ozma of Oz*, Dorothy Gale survives a shipwreck thanks to a poultry crate and a smart and savvy hen named Billina that guides the innocent Kansas girl through a strange land.

This view of the bird as a sagacious creature went all the way back to Aesop, but it began to erode in the wake of World War I and the rise of mass poultry production. As Americans left the farm, moved to cities, and bought refrigerators, live chickens began to slowly vanish from daily life, and attitudes toward the bird hardened. "Get away, yuh dumb cluck!" yells a character in Edmund Wilson's 1929 novel,

I Thought of Daisy. The term *birdbrain* first turns up in 1936, and the insults *to chicken out* and *chickenshit* were first used during World War II. *Chicken Little* made its screen debut in 1943, with hysterical fowl standing in for Germans manipulated by Hitler. The only thing dumber than a chicken was a geek, the sideshow freak who would bite the head off a chicken, a term popularized in a 1946 novel and 1947 film noir starring Tyrone Power.

Three-quarters of a century after the creature became the butt of jokes, scientists are discovering that our species shares a surprising number of traits with the humble chicken. The Italian neuroscientist Giorgio Vallortigara recently demonstrated, for example, that just-born chicks are natural mathematicians. They can track little plastic balls appearing and disappearing behind screens, even when he tried to fool them by moving some balls to another screen. Humans generally fail at the task until they are four years old.

And chicks can do more than add and subtract. They can understand geometry, recognize faces, retain memories, and make logical deductions that Vallortigara insists exceed the capabilities of some of his graduate students. Other neuroscientists are finding that chickens practice self-control, alter their message to fit the receiver, and, in some instances, can feel empathy. Some of these cognitive abilities equal or surpass those of assorted primates, and it is possible that the chicken possesses a primitive self-consciousness.

Vallortigara's lab is in the cellar of a sixteenth-century convent in Rovereto, an Italian town in the foothills of the Alps. A dapper man in a light-blue shirt and silk tie, Vallortigara was born here in the decade after World War II, when Italy was poor and flocks were critical to survival. "Not to have chickens was not to have eggs," he says, and that often meant going hungry. As a child, he grew curious about how animals perceived the world.

The idea that animals have mental abilities similar to those of humans has been controversial since the seventeenth-century French philosopher René Descartes declared that they lack minds, reason, and a soul. Animals communicate anger, fear, or hunger with sounds, he noted, but they don't speak and therefore lack the inner voice that is the very basis of human thought. His famous "I think therefore I am" might be better stated as "I speak therefore I am." Animals might feel pain or pleasure—"I deny sensation to no animal," he wrote—but they lack that human quality of greater awareness or cognition. Philosophers, scientists, religious figures, and animal activists have been arguing this point ever since.

Neuroscientists like Vallortigara gather hard data on animal perception. They have learned that chickens see the world in far greater depth and detail than we do. Mammals began as nocturnal creatures to avoid predators like day-loving dinosaurs, while birds preferred the light and therefore have far superior color vision. A red jungle fowl boasts rich reds and blues and greens in its feathers, but the bird perceives a dazzling combination of colors extending into the ultraviolet and beyond human vision. Chickens also use each eye for separate purposes, allowing them to focus on one object—such as potential food—while keeping another eye peeled for predators. This accounts for the odd and jerky head movements of a hen or rooster.

The chicken's excellent visual system was once thought to be offset by its inability to smell. Recently, a research team studied a domestic flock that did not react when presented with elephant and antelope dung, but grew alert and stopped eating when placed near the droppings of wild dogs and tigers. Like humans, chickens may rely more on their eyes than their noses, but they can catch a whiff of danger. The bird also recalls faces of humans and chickens, and can respond to that individual based on a previous experience. Seeing a favorite hen, for example, will cause a rooster's sperm production to suddenly increase.

Scientists once scoffed at the idea that fowl have a sophisticated method of communication. "Even if chickens had a grammar," the cognitive psychologist David Premack wrote in the 1970s, "they would have nothing interesting to say." Since then, a German linguist has concluded that all fowl have about thirty distinct sounds matched to specific behaviors. For example, the bird uses separate calls to signal whether a predator is approaching by land or by air.

Vallortigara studies chickens primarily because they are inexpensive, hardy, and easy to maintain. Most birds, like mammals, require extensive care of their young, while chickens emerge from their eggs largely capable of fending for themselves and ready to take part in experiments before the external environment influences their behavior. In the old convent cellar, a team of seven PhDs and assorted graduate students in orange shoes and white coats bustle through the narrow and brightly lit hallways, which nevertheless have the air of a dungeon. Vallortigara takes me first to a dark, warm room filled with fertilized eggs that will soon serve as test subjects.

The experiments focus on filial imprinting, in which newly hatched birds attach themselves to the first moving object they see. The researchers show a chick a particular object such as a red cylinder, place the animal in a transparent pen, and hide the cylinder behind one of two opaque screens. The transparent pen then is covered with a screen, and after a delay of up to one minute, they let the chick go to the screen of its choice. The bird finds the imprinted object on the first try, demonstrating a well-developed memory. In another experiment, the cylinder is hidden behind a screen that covers the entire object, while the neighboring screen has a different height or width that would reveal a portion of the object. The bird invariably picks the one disguising the cylinder, a sign of what Vallortigara calls "an intuitive physics."

Chicks also are capable of adding and subtracting. A young bird presented with an identical cylinder that disappears behind one screen and several that disappear behind the other screen will go to the screen with the larger number of cylinders. If the researchers move a cylinder from one screen to the other, so that the second screen holds more objects, the chick will go to the screen with the most objects. In another experiment, six identical containers are set up in an arc, all of which are the same distance away from the bird but only one of which has feed, and the chick is allowed to find which one has the food. Switch the food container to another position, and the chick will still choose the correct one.

Vallortigara and colleagues also recently determined that chickens have brains with distinct right and left lobes, an attribute long thought to be solely human. Our left hemisphere controls language, the tool that Descartes believed set us apart from other beings, while our right hemisphere orients us to the people and objects around us. A developing chick's left eye is covered while the right eye faces the shell of the egg. Exposing the right eye of the embryo to light in its last three days in the egg weakens the bird's visual processing. Once hatched and faced with pebbles mixed with grain, the left side of the normal chicken brain judges what is pebble and what is grain and then pecks only the food, while the altered chick cannot distinguish between the two sets of objects.

Chickens make use of the different halves of their brains for specialized tasks, and, Vallortigara suggests, they can distinguish between animate and inanimate objects. In another experiment, chicks are shown random points of light as well as points of light that mimic a walking hen, cat, or other animal. Chicks invariably prefer the lights that imitate the motion of a biological creature, though not necessarily the hen-like image itself. Normal two-day-old human babies also make this distinction, but many autistic children and adolescents cannot. Vallortigara's team is exploring whether the disorder is linked to this instinctual ability to interpret biological motion. By pinpointing the genes operating in chicks during biological motion cognition, Vallortigara hopes to understand the mechanisms that might go awry in autistics, a first step in treating the disorder.

In 2012, an Australian philosopher reviewed the scientific literature and concluded "it appears likely that one familiar species, the chicken . . . exhibits primitive self-consciousness." As a result, he added, "chickens warrant a degree of moral standing that falls short of that enjoyed by persons, but which exceeds the minimal standing of merely conscious entities." They not only can feel sensation, as Descartes knew, but they may know that they exist and, therefore, that they suffer.

While Vallortigara studies what makes chicks tick, other scientists are probing how chickens feel about living conditions in today's industrial poultry business. Animal-rights activists and companies that sell eggs are locked in a battle over how hens are treated, and the research results could determine the fate of billions of birds for decades to come.

The father of today's egg-laying hen was animal lover, some-time vegetarian, and pacifist Henry Wallace, an Iowan who served in President Franklin Roosevelt's cabinet and as his vice president during World War II. A farmer deeply concerned with rampant rural hunger during the Depression, Wallace was convinced that making a more productive chicken could help banish poverty. He founded the Hi-Bred Corn Company in 1926, and his son began to breed commercial hybrid layers in 1936. By the middle of World War II, the business was selling laying Leghorns. Its successor, Hy-Line International, is now the world's largest single producer of egg-laying hens. White Leghorns, which produce white eggs, and Rhode Island Reds, which make brown eggs, are the world's two primary laying varieties. American hatcheries churn out 500 million egg-laying hens a year, and Wallace's home state of Iowa produces nearly twice as many eggs as any other U.S. state. The hens lay more than 75 billion eggs a year.

Nine out of ten laying hens in the United States live in wire enclosures called battery cages. Eight birds may be assigned to each cage, without the space to open their wings. Their heads can poke out far enough to peck at the feed bin that runs alongside cages stacked in several tiers, and their waste falls through the wires onto a conveyer belt below. There is no place to roost, dust bathe, or lay eggs in private, three behaviors common to all hens. Vicious pecking, avian hysteria, mysterious deaths, and even cannibalism are often the result. Birds are debeaked, without anesthesia, to limit injuries as they churn out eggs. Disturbing conditions like fatty liver and swollen head syndrome, mouth ulcers, and feet deformities are common. The noise is deafening, the air filled with ammonia, and the animals have a crazed look.

"In no way can these living conditions meet the demands of a complex nervous system designed to form a multitude of memories and make complex decisions," concludes one poultry researcher. "A gallinaceous madhouse," is how a shaken Texas naturalist summed up a typical laying shed. Wallace's vision of affordable eggs for U.S. consumers came true, but there is growing consumer unease with the methods used to produce the billions of eggs laid in the United States annually.

Chickens are excluded from U.S. laws regulating humane treatment of animals raised as food, and there are no international regulations. Banned in the European Union, battery cages are being phased out in a growing number of U.S. states. Costco and Walmart sell only cage-free eggs under their private brands, while Burger King and Subway, among other food purveyors, are committing to non-battery eggs as well. Yet scientists know little about the impact of various sorts of confinement on chickens. A label designating eggs produced in a cage-free environment might trigger images of happy hens tripping across a sunny meadow, but the vast majority of these birds are raised like broilers on the floor of vast, enclosed sheds or in multistory aviaries and are subject to violence, disease, and neuroses. We might feel better, but do the chickens?

A new $1.8 million research facility at Michigan State University in East Lansing is designed to answer that question. Animal welfare scientist Janice Siegford, who is conducting experiments in the boxy building on the outskirts of campus, is part of a new generation of researchers who want to use science to improve the treatment of chickens while accepting the reality of consumer and industry economics. Siegford worries that the debate about battery cages is driven by emotion or economics rather than data, so she has begun tracking the behavior of laying hens living in three different housing systems.

Siegford, athletic and wearing close-cropped hair, is preparing to take freshman and sophomore students on a tour of the facility when we meet. The neuroscientist began her career experimenting on the spinal cords of Mongolian gerbils to understand how neurons are generated, work that could provide clues to treating human paralysis. Frustrated with the field's tendency to reduce everything to its component parts rather than consider the whole organism, she switched to animal welfare research. "Helping people with paralysis is pretty cool stuff," she says. "But I thought if I was going to use animals in my research, then I would like it to benefit them directly."

When the students arrive, Siegford explains the goals of her research program and challenges them to consider difficult questions. "Our idea is to find out what is good for the chickens so that they have a better quality of life. As you go through, consider the pros and cons of chicken health, your health, and economics, too—there's a lot to think about." In an anteroom we don white suits designed to protect the seventy-two hundred birds from illness and then enter the wide main hall that leads to a dozen separate rooms. Four contain aviaries, tiers of cages with an open area of litter on one side. Eight have cages that contain perches, dust bathing mats, and screened-off nest boxes—"enriched" cages in poultry parlance. These two systems are the leading alternatives to broiler-shed or battery-cage models.

Siegford guides me into the cloying warmth of an aviary bathed in eerie pink light, with three tiers of cages and an open area on one side of the floor housing several hundred White Leghorn hens. Allowing the birds to roam the floor and tread in dung gives them more freedom and increases the odds that harmful microbes could be transmitted to their eggs. The open area is divided into Astroturf, sawdust, straw, and bare concrete so that researchers can determine which environment appeals the most to the hens and test the microbial load of each flooring material. A student group troops into the room. "Why do some have big bald spots?" asks one. "Chickens are not kind and gentle," replies Siegford. "They peck at each other and pull feathers out. When a hen lays an egg, her cloaca pushes out and is bright red. In a sea of white chickens, that is startling and attractive and they may peck at that. And some of the spots come from their rubbing against the cage."

The enriched cage next door gives birds more room for movement than a battery cage, but there is no open floor space where they can gather. Bright-orange curtains close off nest boxes and hens sit on small perches, features absent in battery cages. Another enriched cage includes a greater assortment of perches and nest boxes, as well as small plastic mats designed for dust bathing. Siegford wants to understand what size a nest box must be to assure privacy and what height is optimal for a perch. Getting such small details right is necessary to ensure the birds' comfort and mental health.

Hens are more likely to die in an aviary, fall prey to cannibalism, or suffer from the higher levels of ammonia and dust stirred up by their movement, says Siegford. Gathering eggs is also more difficult for workers, who must use rakes and slide around on their bellies. The birds in the enriched cages, by contrast, appear to have more bald spots around their necks. "This one's seriously feather pecked," she notes as she peers into the cage. "Oooh, that's bad."

The research program is in its early stages, so she can't yet draw firm conclusions, but she's skeptical that the open-floor approach improves overall chicken welfare. "Aviaries let chickens have more freedom," she says, "but I don't think they always use that freedom wisely." Siegford worries that animal-rights organizations, states, and industry will shift to aviaries before it is clear that they are an improvement. She believes that enriched cages may be the better choice for practical as well as welfare reasons, since the industry already is used to managing caged birds.

How much space a bird needs to live comfortably remains a matter of dispute between activists and industry, and Siegford hopes her research will provide clear answers. Every single hen in the facility is weighed, tagged, and given a welfare quality score, data that might reveal when overcrowding triggers increased violence and disease. The point at which chicken's social structure, called the pecking order, breaks down is not yet known, although Siegford estimates that the limit is a community of fifty birds.

Stepping back into the hall, she points out a panel regulating water, feed, and air temperature for each room. Her goal is to add a sensor that can alert operators if birds go silent or panic en masse, a feature that could become part of commercial laying operations. Observing a flock's behavior is simpler than attempting to track an individual among thousands, so Siegford and an MSU engineer turned to sensing technology used in the military to track the location and status of soldiers in the field using electronic harnesses. These are too large and expensive to use on chickens, so the pair designed a miniature sensor and went to a pet store and bought a hundred Chihuahua harnesses—"the clerk thought we were crazy"—that could be slipped over a bird. Such a device ultimately could be used to spot depressed appetite or increased pecking among individuals, signaling health or behavioral trouble before they reach epidemic proportions.

More humane environments and better monitoring would benefit layers, but they will not address some of the most disturbing practices in today's industrial agriculture. Layer hatcheries discard all roosters, and there are no U.S. government regulations on how this should be done. No law prevents abuse of the unwanted animals. Laying operations frequently limit food or starve aging hens to speed molting so that they will resume laying more rapidly.

Within a year or two, a typical laying hen is spent. "We selected chickens to lay eggs at the expense of their own bodies," says Siegford. "You won't see a physically healthy hen at the end of her laying cycle no matter what nirvana you put her in. Her body is driving her to lay eggs." No matter how many perches, dust-bathing mats, or open space are provided to hens, the genetics of the bird works against its welfare. And once their laying days are over, there are no laws governing their destruction, and the hens have almost no economic value except as fertilizer or pet food.

As we step outside and the students depart, Siegford's faith that science will build a bridge to better chicken welfare wavers. The scale of commercial layer operations boggles her mind. "When you look down the row at a laying house, you can't see to the end. How many birds are there, and how do you keep track of their welfare? What bothers me is that we are so disconnected from these living beings, and our role as caretakers, in the name of efficiency and cheap food." And chickens, she adds, evoke less interest and empathy than cows or pigs. "They are like background noise. But not if you really look at them. They use the sun to tell time, their social communication is quite complex, and they are one of the few animals besides primates to alter their message to suit their audience."

The campus of Jomo Kenyatta University of Agriculture and Technology on the outskirts of Nairobi is a lush and shady retreat from the noise and traffic of the Kenyan capital. Sheila Ommeh, who is tall even without her black strap-on heels, meets me at the main gate. The molecular biologist is Africa's Henry Wallace, a single-minded poultry promoter on a mission to give chickens a starring role in the development of her country and continent.

"Chickens may be small, but if they are properly harnessed, they can have a big impact," she says. "There's great potential to benefit farmers across Africa." Only one Kenyan chicken in ten is raised on an industrial-scale farm, but indigenous chickens are too expensive for most city dwellers and rural people. A rooster can cost twenty dollars, a large sum in a country where the average annual salary is under a thousand dollars and unemployment runs at 40 percent. Ommeh put together a joint effort with the university, the National Museums, the Kenya Wildlife Service, and the Ministry of Livestock Development to transform the expensive and neglected bird into a source of protein and income for poor Kenyans, a model she hopes to spread across Africa.

A microbiology class with a dozen graduate students is in session in her lab, so Ommeh takes me across the dusty road to an empty chicken house. Previous researchers raised a layer with indigenous genes developed by a French company, but they proved too expensive and require vaccines and special feed unavailable in most rural areas of the country. Ommeh intends to replace those birds with tougher and cheaper local varieties that lack the egg-laying capacities of industrial birds but are better adapted to the realities of Kenyan farming.

"The chickens will have plenty of room to play around in the sunshine," she says, gesturing toward the field adjacent to the coop. Ommeh grew up in western Kenya, in a village on the slopes of fourteen-thousand-foot Mount Elgon, and as a young biologist she landed a position in Nairobi's prestigious International Livestock Research Institute, which emphasizes large livestock like cattle. "I came from a community of chickens and my grandmother watched helplessly as her chickens died of disease. She would say to me, 'You are a scientist but you don't help us!' I realized I would benefit my community more by working on chickens." Unable to convince the institute to focus on poultry, in 2011 she left her prestigious post at the university and began to build support among academics, development specialists, and politicians.

In the 1970s, Western aid agencies introduced the Rhode Island

Red and other varieties of industrial chickens that quickly died off, but not before mixing with local birds. The resulting hybrids lacked tolerance for the tropical diseases and local flocks were decimated. Ommeh is searching for those few Kenyan birds with enough indigenous genes to survive dry spells and avian illnesses endemic to the country. "Making sure they are disease and drought tolerant is more important than talking about their meat productivity," she says. "And we have to move fast or it will be too late."

Ommeh targeted Lamu on the Indian Ocean coast, near the border with Somalia, to seek out the last of the purebred Kenyan chickens. The country's oldest continuously settled city, Lamu is a sleepy backwater that once was a bustling port filled with African, Indian, and Southeast Asian merchants. She discovered a large local multicolored chicken called the Kuchi kept as an ornamental pet and gamecock. She also explored the remote area around Lake Turkana in the northwest, near Ethiopia and South Sudan, where chicken soup and eggs are an important part of the diet of tribespeople who primarily herd cattle. Small but extremely hardy, the white bird thrives in a stark landscape where temperatures regularly soar above 108 degrees.

At the lab, Ommeh holds up her field kit for gathering chicken samples—a Samsung Android phone. "This is better than a laptop—you just need batteries and you are ready to go." In Lamu and Turkana she uses the phone to input vital information like the coordinates of the sampling site, the bird's laying capacity, and its adaptive traits. She also drips blood samples on chemically treated cards that require no refrigeration and scans their individual bar codes to link the sample with the phone data.

Back in the lab, she analyzes the genetic makeup of the chicken blood to pinpoint the mechanisms behind the birds' tolerance to disease and extreme environmental conditions. The result, Ommeh hopes, will be chickens that are easy to care for, inexpensive, and adaptable to rural Kenya. She smiles, returning the phone to her pocket. "My grandmother would be very happy that my science is doing something practical."

13.

A Last Cause

How long have humans and nonhumans been carrying on
this way?

      —Alice Walker, *The Chicken Chronicles*

Nguyen Dong Chung's village sits amid the electric-green rice
fields of Vietnam's fertile Red River valley. A slight, middle-aged man
in a white button-down shirt and dark trousers, Chung is the protector of the Ho chicken. Before the revolutions and wars that rocked his
country in the twentieth century, the finest pair of Ho was presented
to the king and queen in nearby Hanoi during the Chinese New Year
festival. On special occasions, a Ho is still slaughtered, cooked, and
presented to the ancestors, and then eaten by the descendants. "My
grandfather and his grandfather raised Ho," he says through a translator. "We are proud of them. We think they are very important."

    Chung, who has maintained his family's profitable business sell-
ing Ho chickens, draws customers from around the Hanoi area. He
guides me through a warren of alleys to a series of coops where he
raises the bird, an ungainly creature with huge feet supporting a large,
thin frame with dark feathers and red skin topped with a rose comb.
"They are big and beautiful and their meat is tasty," Chung says. The

eggs go for a pricey three dollars each and a chick for five dollars. At a nearby shop, his Ho gamecocks are for sale at two hundred to three hundred dollars apiece.

The Ho are among the last of the world's ancient chicken varieties. Western heritage breeds and the modern industrial chicken are descendants of the nineteenth-century British and American hen fevers. Village fowl like the Ho, by contrast, date back many centuries and even millennia. Vietnam alone has sixteen distinct chicken breeds that account for three out of every four chickens raised in the country. As industrial broilers and layers proliferate around the world, these village birds are quietly disappearing. That fact worries the Chinese biologist Han Jianlin, who collects genetic data on chickens like the Ho and introduced me to Chung. "The system is changing fast, and the old varieties may vanish with the older generation," he says as we return to the smog of Hanoi. As they vanish, diverse and useful traits cultivated by local breeders over thousands of chicken generations will be lost.

Like Ommeh, Han wants to preserve traits that might benefit poor farmers around the world. Since 2001, he and his colleagues have amassed nearly thirty thousand samples of village chickens from Angola to the Philippines. A broad-shouldered man with thick hair and perfect English, he has a state-of-the-art biology laboratory in Beijing but spends much of his time in the field, visiting breeders like Chung and monitoring efforts to create birds well suited for a particular climate, ecology, and palate. I accompany him next to the National Institute of Animal Husbandry in a southern suburb of Hanoi, where he checks up on some of his latest experiments. We don rubber boots and white coats at the entrance and walk among rows of long concrete coops with metal roofs that house Vietnam's poultry laboratory. His goal is to create hybrids that are tasty, resistant to local diseases, and more productive than existing varieties. If a new strain seems promising, the chickens are distributed to villagers and farmers around the country.

Han's passion for village chickens took root as he was growing up in China's western province of Gansu in the 1960s, when it was illegal for an individual to own a chicken and millions perished of starvation under the communist government of Chairman Mao Zedong. When those restrictions were lifted in the late 1970s after Mao's death, Han raised a flock, collected the eggs, hid them from his hungry sister, and periodically sold them at a market to buy books and pencils for school. That led to college, a PhD in biology, and a certainty that poultry offers enormous advantages over larger livestock.

One-third of South Asians and more than half of all sub-Saharan Africans suffer from malnutrition or undernutrition, mostly in rural areas. Dependence on grains like rice can make pregnant women and young children more vulnerable to disease. Chicken meat and eggs offer a high dose of protein as well as vitamins and minerals like lysine, threonine, and other important amino acids that lower the risk of macular degeneration and cataracts. Village chickens can reduce child mortality, improve the health of pregnant mothers, and contribute to public health by eating insects that spread disease. They don't compete with humans for food, as pigs can, and their care requires little infrastructure.

As Han knows from personal experience, the bird and its eggs can help families pay for school fees and supplies otherwise beyond their means. When he discovered that chickens are the world's most studied animal, but that village birds were largely ignored by industry and academic poultry specialists, he set out to find, catalog, and study these forgotten fowl.

The next morning, I join Han and Le Thi Thuy, a senior scientist in animal husbandry at the Hanoi institute, on an expedition to the country's mountainous northwest to visit research stations and perhaps encounter a red jungle fowl in its natural habitat. The remote and rural region bordering Laos and China is renowned for its tasty chicken varieties. The following evening, at a popular restaurant in a town high in the mountains, we sample three kinds of chicken, each prepared as medallions that include skin, bone, fat, gristle, and meat, and served with the cooked comb. One dish is honey-colored, another metallic gray, and the third black. "Which do you prefer?" Han asks as Le digs into the feast. I confess that the darker the meat, the less palatable I find it. He laughs and explains that the oilier and darker the meat, the more it is prized.

In Vietnam, chicken is not simply a meal. The organs, bones, blood, and feathers as well as the meat of what is known as the Hmong chicken are black, and the bird is famous in north Vietnam for its ability to enhance a person's vitality, liven up a sluggish sex drive, and remedy heart disease. The Tre chicken in the country's south is a favorite gamecock, fierce in the ring. The Tre, bred in the central part of the country, is popular as a dwarf breed.

The country's rapid development and fast-growing population, which tripled in the past fifty years and soon will top 100 million, is changing the ancient relationship between Vietnamese and chickens. The number of birds raised here annually doubled from 1995 to 2010, to 200 million. Local varieties still dominate, but they are losing their market share to industrial chickens as young, urban Vietnamese frequent fast-food outlets like KFC, which recently opened its thousandth restaurant in the country.

After China, Saudi Arabia, and Singapore, Vietnam is the world's largest importer of industrial chicken. Even with massive imports, however, the country appears to be eating itself up. Han points out the window of our truck at bright orange stripes that crease the northwest mountains. He says that in order to feed the factory-farmed pigs and chickens that are a growing part of the Vietnamese diet, farmers cut down trees to plant corn on steep slopes, but after a few harvests, the thin soil washes away in the tropical downpours, leaving ugly orange gullies that the forest cannot easily reclaim.

The large and pale-breasted industrial bird is even showing up in rural areas far removed from Hanoi's supermarkets. In one market town near the border with Laos, two plucked industrial chickens lay at the far end of one market vendor's table. Though cheaper than the half-dozen other local breeds she offered, they weren't selling. "No one really likes them," the vendor confided. "They don't have any flavor." Yet the bird's very presence marks change, a change that Han sees sweeping other rural areas from Africa to India.

The threat of avian influenza in Southeast Asia may be as much or more of a threat to the traditional village chicken than KFC. Italians first described the illness in the wake of Chinese chicken imports in the 1870s, but few scientists suspected until recently that deadly viruses could jump from chickens to humans. Recent research suggests that the 1918 pandemic that killed 50 million people and sickened about one-third of the entire human population may have begun in chickens and then moved to pigs before infecting humans. In the late 1990s, a strain named H5N1 afflicting Hong Kong water-fowl jumped species and killed a half-dozen people. All 1.6 million of the city-state's live poultry were destroyed to halt its spread. A half-dozen years later, birds began to die in northern Thailand and southern China. "From the tips of their combs to the claws on their feet, the birds literally *melt*," one horrified Western scientist reported. The virus again leaped species. Half of the hundred people infected died—a shockingly high rate—and tens of millions of birds across the region were killed to contain the epidemic, with Vietnam particularly hard-hit.

Such threats boost industrial-chicken enterprises at the expense of traditional backyard chickens and live poultry markets. Growing birds in enclosed and isolated facilities that strictly limit their contact with anything on the outside, such as ducks, pigs, and humans, reduces the risk of deadly species-to-species transmission. That fact encourages governments to back large and centralized poultry operations in the name of public health. In 2004, Thai prime minister Thaksin Shinawatra threatened to ban villagers from raising hens and gamecocks, a policy supported by large chicken producers like Bangkok-based Charoen Pokphand, which employs more than a hundred thousand people and accounts for about 10 percent of Thailand's gross domestic product. A massive culling of village chickens infuriated small farmers across the country. Many of them received only small sums for expensive game fowl that were slaughtered. Some researchers suspect that the fast-growing industrial-poultry business has also played a role in the crisis, since the virus spread at the same time that the industry began to expand.

In 2013, a new strain of the H7N9 virus killed more than one in three of the 135 people infected in China. Most of those victims had contact with live poultry, so the infections likely came directly from chickens. But scientists discovered that the strain can rapidly mutate, and fear that it could find a way to pass between humans. Some researchers argue that the best solution is to close China's live-poultry markets. This radical approach worked in Shanghai, Nanjing, Hangzhou, and Huzhou, where the spread of infection stopped as a result, according to a study in the medical journal *The Lancet*. Other researchers say that disinfecting poultry markets once a week could reduce the spread of the virus without undoing a traditional part of the Chinese social fabric. The combination of economics and public-health concerns threaten the local birds and markets that have been part of Southeast Asian village life for millennia.

Han sees the march of the industrial chicken as unfortunate but unstoppable. He recently returned to his home village in western China and was shocked to find no chickens or pigs, since local families now buy their meat from a supermarket. He is even doubtful that the Ho cultivated by Chung outside Hanoi will remain economically viable as the price for imported chickens drops and tastes change. Some biologists fear that humanity's dependence on a few breeds of chickens exposes a critical part our food supply to disaster should an epidemic wipe out a large percentage of, say, the Rhode Island Red or the White Leghorn.

Han is skeptical of such a doomsday scenario. Although he is among the strongest advocates for preserving vanishing village varieties, he says that the industrial bird retains a high degree of genetic diversity that makes such a calamity unlikely. Like William Beebe, he has faith in the chicken's astonishing ability to adapt to the needs of humans, if not in the humans' ability to make wise choices.

Chickens arrived on the human scene as humans began to grow and eat grain. But most grains like corn lack important nutrients, particularly amino acids such as lysine and threonine. Chicken meat and eggs are rich in these essential amino acids. "This feature of chicken metabolism—the ability of the hen to lay down in her egg amino acids that man cannot synthesize—has played an important role in the evolution of both species," two academics write. "The presence of only the occasional egg in the diet could well change a marginal meal into one that is nutritionally satisfactory."

If humans continue to eat meat in the twenty-first century, then industrial chicken is a better option than pork or beef. Chickens require less land, water, and energy inputs than pigs or cows. Less than two pounds of feed produces a pound of meat, a fraction of what other animals require. Only farmed salmon can beat that ratio. Meat production accounts for nearly 20 percent of the world's annual greenhouse gases pumped into the atmosphere and more than 80 percent of all the greenhouse gases produced by agriculture. Pound for pound, chicken releases only one-tenth the greenhouse gases of red meat such as hamburger.

The British economist Thomas Malthus warned two centuries ago that population growth would overwhelm our ability to produce food. He could never have foreseen the introduction of exotic Asian fowl, advances in genetics, and the rise of cheap energy and large corporations that transformed the scrawny bird of his day into a mass-produced and inexpensive commodity. A billion chickens weighing more than 12 million tons cross national boundaries each year. Dutch chickens sizzle on the silver platters of Kuwaiti princes, Angolan tribespeople barter in local markets for Arkansas birds, and Brazilian fowl are for sale in Beijing superstores. Global poultry exports expanded more than a quarter between 2008 and 2013, with sub-Saharan Africa and the Middle East accounting for most of that increase. Ghana imported more than two hundred thousand tons of frozen chicken in 2012, three times what it bought a decade before.

Americans' appetite for chicken, which tops that of all other major nationalities, continues to grow. U.S. consumers eat four times the world average. Mexicans are the leading egg eaters, consuming more than four hundred annually, almost three times the global average. In China, people eat twenty-two pounds of chicken a year—only about one-fourth of the American appetite—but that figure is climbing annually and the country is the world's largest market for bird imports. In 2012, for the first time, more chicken was consumed in China than in the United States. It is a stunning change from the days of Han's youth, when poultry production was almost nil. Fujian Sunner Development Co. employs ten thousand people in its vertically integrated poultry operation in China, and its chairman Fu Guangming is a billionaire. Tyson Foods intends to build ninety farms in China in coming years, each with more than three hundred thousand birds at a time, as part of an effort to capture a share in what is now the largest single market for chicken, and other U.S. companies like Cargill are also rushing to build their own farms and processing plants. Little wonder that Han's village now lacks chickens. In 2014, the Chinese government announced it would move 100 million more people to cities from the country by the end of the decade, so that 60 percent of the population would be urban.

Malthus also could not have foreseen the daunting environmental, labor, health, and animal-welfare issues that bedevil our radical new relationship with the bird. Today's poultry industry, like the proliferation of megacities and human-induced climate change, is an experiment on a scale and of a scope never before attempted. Polluted waterways, dangerous conditions for workers, food safety concerns, and appalling animal-welfare issues remain largely in the shadows of this vibrant international trade. The companies that dominate the industry tend to be large and politically powerful and capable of blocking or watering down government regulations that could increase their costs.

Someday we may lose our appetite for chicken. Alternatives are appearing in European and American markets, such as chickenlike tofu or mushrooms that mimic the bland taste of industrial chicken. By heating, cooling, and pressurizing vegetable protein, the owner of the California company Beyond Meat promises to create an inexpensive chickenlike enchilada free of antibiotics, arsenic, and avian flu. Its "chicken-free strips" are now sold at the Whole Foods supermarket chain. Another California startup, Hampton Creek, wants to replace the egg with a vegan option, thanks to funding from PayPal cofounder Peter Thiel as well as Microsoft founder Bill Gates. Its first product, Just Mayo, is now on shelves. While many of these alternatives taste as bland as the food they aim to replace, they are poised to capture a growing share of the American and European market.

Yet with the growth of megacities like Hanoi and Nairobi, the advancing middle class of India and South America and the retreat of village varieties, the chicken seems locked in as our fast-food staple and urban necessity for decades to come. The royal pet, the sacred symbol of the sun and herald of resurrection, the bird that cleansed us from sin and provided us with our models for courage and self-sacrifice is quickly turning into a vital foodstuff. The chicken in China, for example, has long represented the live virtues of politeness, martial arts, bravery, benevolence, and faith, but its primary role today is feeding the country's 150 cities with populations in excess of 1 million—a number likely to double by 2030.

More humane genetics, treatment, and living conditions could roll back the worst abuses against our companion species without unduly interfering with the flow of cheap animal protein to our cities. Egg cartons, for example, could specify in simple images the conditions of the hens. The cartons that I saw in Bourg-en-Bresse were clearly marked according to the condition of the hens that laid the eggs, as mandated by French law. Chicken fertilizer could be used intelligently to replenish depleted soils, as Easter Islanders did centuries ago. Labor advocates could follow the approach taken in the textile and electronic industries in countries like Bangladesh and China to expose and then demand improvements to the harsh conditions endured by many poultry workers. Chefs could insist tastier alternatives to the industrial chicken, even if that entails a slightly higher cost. And international guidelines on antibiotic use, feed safety, and salmonella and other bacterial threats could ensure consumers that buying cheap industrial chicken, whatever its origin, is not a health gamble.

Industrial chicken may be an inevitable part of our human future, but we can reshape the way we breed, treat, and slaughter the birds while expanding our choices in restaurants and supermarkets.

The backyard chicken movement in Europe and the United States is one hopeful sign that the word *chicken* can mean more than just a meal choice. "Don't try to convey your enthusiasm for chickens to anyone else," warned E. B. White in the 1940s. That interest is now something to crow about. Though the numbers are small and the trend may prove fleeting, communities have banded together to roll back provisions banning chickens within city limits in dozens of American cities, exposing many people unfamiliar with the bird to its charms and advantages. The growing popularity of chickens—there are now whole sections in magazine stands devoted to poultry—has played a pivotal role in the resurgence of small-scale agriculture across the country. The spike in raising homegrown vegetables followed in the wake of early-twenty-first-century hen fever.

Upscale Neiman Marcus recently offered a hundred-thousand-dollar chicken coop, complete with chandelier and heritage birds, as its ultimate extravagant gift. "We'll hold the future farmer's hand as we teach how to raise a healthy flock, to compost from hen house to garden bed, grow veggies and herbs for their table and flock, sprout legumes, fodder and rotate pasture trays, and experience the reward of permaculture," says the designer.

Just as the Victorian Fancy cooled, so will today's backyard birds fall out of favor, and the number of chickens abandoned in city parks or at animal shelters is on the rise. The movement, however, may also help reorient and revitalize our relationship with the animal. We may pause to confront the problems we've created in relegating the chicken to the role of industrial machine. We may pause, if only for a moment, when deciding what's for dinner. We may remember that the chicken is a living animal with a fascinating and formidable history as well as affordable meat.

Marx and Lenin look sternly from their frames as Le, Han, and I—a trio of former enemies, Vietnamese, Chinese, and American—plot a campaign to find the red jungle fowl in its native habitat of Southeast Asian jungle. Dawn and dusk are the best time to spot the notoriously wary wild bird. The local method is to use another captured fowl to lure it out of its hiding places. "The snare is carried in a basket-like case, which is often fitted with a compartment for the decoy rooster," writes Fay-Cooper Cole, who founded the University of Chicago's anthropology department and studied the Tinguian people of the Philippines in the 1920s. "This is set up so as to enclose a square or triangular space, and a tame rooster is put inside. The crowing of this bird attracts the attention of the wild fowl who comes in to fight. Soon, in the excitement of the combat, one is caught in a noose, and the harder it pulls, the more securely it is held."

The live bait, in this case a red jungle fowl captured in the wild a few weeks before, waits sullenly in a small wire cage in the back of a Russian Jeep parked outside a community hall of a north Vietnamese village.

After a flurry of cell phone calls, we set off in the Jeep as the day draws to a speedy tropical close. On the curvy highway that snakes through a lush valley squeezed by saw-toothed mountains, Han points out swaths of corn planted on steep hillsides that are washing away, corn that feeds the growing number of industrial chickens in Vietnam even as it destroys the habitat of the bird's wild sibling. Along the way we stop to pick up our guide, a member of the Black Thay ethnic group named Lo Van Huong, and follow his single-syllable directions. Turning off the paved road, we bump over a rough path, ford a rocky stream, and pass through a village of wooden houses on stilts. In low gear, we grind our way up a steep track past lumbering carts of newly harvested rice. Tall and dignified Black Thay women wearing black headdresses with bright needlework swoop along on shiny motorbikes. One clutches a dead chicken that swings rhythmically from a handlebar as she glides past.

As the road becomes impossibly steep, we pull into a saucer-shaped valley hemmed in by stone cliffs, where a lone farmer in a conical hat is harvesting rice. The cage strapped to his back, Huong quickly heads up a path that winds through the rock, and I hurry after, while Han and Le wait below. Beside a stream near the mountain's crest, the guide sets the fowl's cage on the ground and we hide in the surrounding thickets. The feathers of the caged bird glow in the fast-fading light, but there is no sign that the bait is attracting the attention of a wild fowl. Night falls and we give up, but return before dawn the next day.

As the sky lightens, Huong expertly builds a screen of vegetation before noiselessly vanishing behind a couple of bushes. Through an opening I can see the trapped red jungle fowl standing erect in his cage. A long half hour passes amid the whir of mosquitoes and the blare of truck horns from the distant highway. Then the caged bird abruptly shakes its feathers, raises its head, and lets loose a surprisingly loud and deep crow, a cock-a-doodle-doo with an extra note on the end.

A wild cock in the vicinity answers, and that call is echoed by another response from a neighboring ridge. Then the bird in the cage lapses into silence for another half hour. With the sun rising fast, the chances for spotting a wild red jungle fowl are rapidly diminishing. Huong reappears without snapping a twig and we pick our way past boulders and cling to vines as we work our way back down to the rice field. When we arrive at the truck, he speaks briefly to Thuy. "There are a lot of jungle fowl here," she translates. "But you make too much noise." I murmur that I'm no William Beebe. Then she adds that while chatting on her cell phone, she saw a fowl fly across the little valley.

After saying good-bye to Huong, we drive deeper into the craggy mountains, and the sun is low and our legs cramped when we arrive at a farm outside of the city of Son La. A bamboo cage containing a red jungle fowl hangs in the courtyard. Nguyen Quir Tuan, a lean farmhand from the Hmong ethnic group, says that this bird came from high in the mountains, beyond the reach of humans and domesticated chickens. He opens the cage and takes the reluctant bird by its feet. Its spurs are razor sharp, longer than my longest finger.

"There are fewer now because the trees are being cut and they are being hunted," says Tuan as he puts the nervous bird back into its cage. "Years ago, when tigers were still prevalent, the fowl inhabited even the valleys," he says, gesturing at the fields around us. Now, he adds, the only wild places are a few isolated mountaintops. Tuan expresses an unmistakable awe of the animal. He says it is smart and secretive and can swiftly die if caged by rushing the bars and breaking its neck. Though small, he says it is tough enough to beat a domesticated rooster in a fight, and can also fly.

As we tour the countryside in subsequent days, captured red jungle fowl hold a place of honor in village homes. In the small community of Chieng Ngan, a local official takes us to visit a farm family with a small coffee plantation and a pet wild bird. "They can live twenty years," he says as we climb the stairs of the plank house on stilts with a low-pitched roof. A young father with a toddler on his hip—not an uncommon sight in this matrilineal culture—greets us. By the open window stands a young male jungle fowl with a string tied to its leg, next to two sliced-off plastic bottles of rice and water. The owner says he keeps the animal for its crow.

In Vietnam, it is illegal to hunt or trap the red jungle fowl, and villagers are either creative or vague in explaining the presence of a wild fowl in their homes. One woman says that her husband came across a jungle fowl egg while searching for mushrooms, and another claims that a wild fowl was caught while fraternizing with her chickens. All say they take particular delight in the bird's crow, which they found more melodious than that of a domesticated rooster. No one would say whether the captured males were used for cockfighting or for a good meal. But after we waved good-bye to the official, Thuy laughed. He had told her that nothing tasted as good as the fowl, and that he ate about twenty a year.

The next morning we visit Son La's outdoor market, where one vendor displayed commercial brown eggs and more expensive local chicken and duck eggs. "Do you have red jungle fowl eggs?" She jumps up from her stool and hovers boldly over my notebook, and I hand her my pen. She scribbles something, saying, "No eggs, but order a jungle fowl." The cost, she adds, would be about a hundred dollars and in my notebook she's written a cell phone number. Red jungle fowl may still live by the thousands across South Asia, but their numbers are diminishing as human populations increase and forest cover shrinks. Most if not all have incorporated genes from the domesticated bird. "There are red jungle fowl that look wild but are not," Han says as we make the long return journey to sprawling Hanoi.

That night, from my hotel in Hanoi, I send pictures of the red jungle fowl to I. Lehr Brisbin, the South Carolina ecologist. "*Way* too calm," he writes back. "Probably not pure stuff." Then he asks me to pluck some feathers and quills, seal them in a Ziploc bag, and bring them back for DNA testing.

Brisbin, the man who saved Gardiner Bump's Indian red jungle fowl from destruction in the 1970s, is enlisting a new generation of researchers to ensure that the remaining Bump birds survive well into the future. At Virginia Tech, biologists recently incubated eggs laid by Leggette Johnson's flock of wild red jungle fowl, the pure descendants of the Bump birds, in order to maintain a flock that will be sustainable for years to come. Brisbin's patient campaign to save the red jungle fowl is finally bearing fruit.

Returning from our excursion to Johnson's Georgia farm, he lays out his dream of repopulating South Asia with what he believes are the last of the truly wild chickens. Restocking forests from India to Vietnam with the Bump birds could halt or even reverse the genetic extinction of the wild chicken. His goal is not to supply game for hunters, nor is he convinced that preserving the wild fowl will benefit future poultry breeders. His reason has nothing, really, to do with ecology, breeding stock, or the sanctity of wildlife. "Most people's experience with chicken is in a shrink-wrapped package at the grocery store," Brisbin grouses as we jostle down the rough track of his driveway. "Most people don't even see it as a bird."

By returning a pure red jungle fowl, in all of its skittishness and heightened senses, to the forests and jungle from whence it came, his goal is simple. The act would pay homage to an animal that has proven itself as our most steadfast and versatile companion. "It's the last cause I want to take on," he says as we pull up at the front door and he reaches into the back for his cane and the bag with the dead squirrel that he collected on the highway. Brisbin wants to save the red jungle fowl not for science, industry, or even future generations. He wants to save the bird as a way to say thank you.

ACKNOWLEDGMENTS

The name on the cover does not reflect the many people unmentioned in the text who made this book possible, including Jenny Cook, David Amber, Tom Hunter Aryati, Wayan Ariati, Psyche Williams-Forson, Tristana Brizzi, Tomas Condon, Rudy Ballentine, Collin Brown, Mark Fleming, Ed Rihacek, Eduardo Montero, Paul Farago, Nathan Lilly, and all the staff at Tod's, the people of AMC, my agent Ethan Bassoff, editor Leslie Meredith, Jessica Chin, Stephanie Evans Biggins, and Mahan Kalpa Khalsa, who hatched the idea. For those who patiently shared their experiences and knowledge of all things chicken, I am deeply grateful.

NOTES

INTRODUCTION

1 *Pope Francis I: Revista,* "Bergoglio: El Cardenal Que No le Teme al Poder," July 26, 2009, accessed March 25, 2014, http://www.lanacion.com.ar/1153060-bergoglio-el cardenal que no le teme al poder.

1 *In Antarctica, chickens are taboo*: "Introduction of Non-Native Species in the Antarctic Treaty Area: An Increasing Problem" (paper presented to the XXII ATCM, Tromso, Norway, May 1998), World Conservation Union.

2 *"Both the jayhawk and the man"*: Julian Simon, ed., *The Economics of Population: Classic Writings* (New Brunswick, NJ: Transaction Publishers, 1998), 110.

3 *Jared Diamond, author of the bestseller* Guns, Germs, and Steel: Jared M. Diamond, *Guns, Germs, and Steel: The Fates of Human Societies* (New York: W. W. Norton & Company, 1999), 158.

3 *"Chickens do not always enjoy an honorable position"*: E. B. White and Martha White, *In the Words of E. B. White: Quotations from America's Most Companionable of Writers* (Ithaca, NY: Cornell University Press, 2011), 77.

3 *Though Susan Orlean declared the chicken*: Susan Orlean, "The It Bird," *New Yorker,* September 28, 2009.

3 *In 2012, as the cost of eggs shot up in Mexico City*: William Booth, "The Great Egg Crisis Hits Mexico," *Washington Post,* September 5, 2012.

3 *"They are eating pigeon and chicken"*: David D. Kirkpatrick and David E. Sanger, "A Tunisian-Egyptian Link That Shook Arab History," *New York Times,* February 13, 2011.

3 *When poultry prices tripled in Iran*: Reuters, "Iran's Chicken Crisis Is Simmering Political Issue," July 22, 2012.

4 *"A commodity appears at first sight"*: Nicholas Mirzoeff, *The Visual Culture Reader* (London: Routledge, 2002), 122.

6 *"Everything forgets"*: George Steiner, *George Steiner: A Reader* (New York: Oxford University Press, 1984), 219.

1. Nature's Mr. Potato Head

7 *On a chilly dawn in a damp upland forest*: William Beebe, *A Monograph of the Pheasants* (London: published under the Auspices of the New York Zoological Society by Witherby & Co., 1918–1922), 172.

9 *"Those birds which have been pointed out"*: Edmund Saul Dixon, *Ornamental and Domestic Poultry: Their History and Management* (London: At the Office of the Gardeners' Chronicle, 1850), 80.

9 *The wolf that became the dog*: Melinda A. Zeder, "Pathways to Animal Domestication," in BoneCommons, Item #1838, 2012, accessed May 15, 2014, http://alexandria
archive.org/bonecommons/items/show/1838.

10 *"Members of this most beautiful"*: William Beebe, *A Monograph of the Pheasants*, vii.

10 *While other American zoos kept birds in small pens*: Kelly Enright, *The Maximum
of Wilderness: Naturalists & The Image of the Jungle in American Culture*, Rutgers
University dissertation, 2009, 130.

11 *"Boredom is immoral"*: Henry Fairfield Osborn Jr., "My Most Unforgettable Character," *Reader's Digest*, July 1968, 93.

11 *"Everywhere they are trapped, snared"*: William Beebe, *A Monograph of the Pheasants*,
2:34.

11 *Beebe was particularly taken*: Ibid., Plate XL.

12 *"It is seldom that I have seen"*: Ibid., 179.

13 *Beebe concludes that the red jungle fowl*: Ibid., 191.

13 *He was writing at the dawn of genetics*: Thomas Hunt Morgan, "Chromosomes and
Associative Inheritance," *Science* 34, no. 880 (November 10, 1911): 638.

13 *The fowl's unusual plasticity*: William Beebe, *A Monograph of the Pheasants*, 2:191.

13 *In 2004, a huge international team of scientists*: Genome Sequencing Center, Washington University School of Medicine, "Sequence and Comparative Analysis of the
Chicken Genome Provide Unique Perspectives on Vertebrate Evolution," *Nature* 432,
no. 7018 (December 9, 2004): 695.

14 *Chickens skip the eclipse plumage phase*: William Beebe, *A Monograph of the Pheasants*, 2:209.

15 *Early monarchs in the ancient Near East*: Jeff Sypeck, *Becoming Charlemagne: Europe,
Baghdad, and the Empires of A.D. 800* (New York: Ecco, 2006), 161.

15 *By the Great Depression*: Dian Olson Belanger and Adrian Kinnane, *Managing American Wildlife* (Rockville, MD: Montrose Press, 2002), 45.

15 *In 1937, President Franklin Roosevelt*: Federal Aid in Wildlife Restoration Act, 16 USC
669–669i, 50 Stat. 917 (1937).

15 *"American wildlife management officials now are facing"*: The Thirty-Eighth Convention of the International Association of Game, Fish and Conservation Commissioners:
September 15, 16, and 17, 1948, Haddon Hall Hotel, Atlantic City, New Jersey, International Association of Game, Fish and Conservation Commissioners (Washington,
DC: The Association, 1949), 138.

16 *Bump and his wife, Janet, set out*: "Interior scientist and wife search for foreign game
birds," news release, U.S. Department of Interior, Department Information Service,
April 29, 1949.

16 *In 1959, the Bumps rented*: "Reports on the Foreign Game Introduction Program,"
U.S. Department of Interior, 1960.

16 *Undeterred, Bump went on*: G. Bumps, "Field report of foreign game introduction
program activities, Reports 6–8," Branch of Wildlife Research, Bureau of Sport Fish-
eries and Wildlife, Washington, D.C., 1960.

16 *It was, he wrote,*: "Reports on the Foreign Game Introduction Program," U.S. Department of Interior, 1960.

17 *"Hold on, I have to get"*: I. Lehr Brisbin Jr., interview by Andrew Lawler, 2012.

17 *In 1959, Pan Am inaugurated*: "Pan Am Firsts," Pan Am Historical Foundation, July 26,
2000, accessed March 25, 2014, http://www.panamair.org/OLDSITE/History/firsts.htm.

18 *In early 1970, as the nation*: Brisbin, interview.

19 *He published his findings*: I. Lehr Brisbin Jr., "Response of Broiler Chicks to Gamma-Radiation Exposures: Changes in Early Growth Parameters," *Radiation Research* 39, no. 1 (July 1969): 36–44.

21 *Even Beebe's New York Zoological Park*: Brisbin, interview.

21 *Brisbin eventually returned*: B. E. Latimer and I. L. Brisbin Jr., "Growth Rates and Their Relationships to Mortalities of Five Breeds of Chickens Following Exposure to Acute Gamma Radiation Stress," *Growth* 51: 411–424.

21 *"I thought, wow, here's a chance"*: I. Lehr Brisbin Jr., Society for the Preservation of Poultry Antiquities, "Concerns for the Genetic Integrity and Conservation Status of the Red Junglefowl," *SPPA Bulletin* 2, no. 3 (1997): 1–2.

22 *Plants also face challenges*: Dorian Fuller, interview by Andrew Lawler, 2013.

22 *Searching through dusty drawers*: A. T. Peterson and I. L. Brisbin Jr., "Genetic Endangerment of Wild Red Junglefowl *Gallus gallus*?" *Bird Conservation International* 9 (1999): 387–394.

23 *In a joint 1999 paper*: Ibid.

23 *"You go in there and"*: Leggette Johnson, interview by Andrew Lawler, 2012.

25 *Darwin explained such extravagance*: Charles Darwin, *The Descent of Man, and Selection in Relation to Sex* (London: John Murray, Albemarle Street, 1871), 38.

26 *He was part of the team*: "Sequence and Comparative Analysis of the Chicken Genome Provide Unique Perspectives on Vertebrate Evolution," 695.

26 *In 2011, he visited*: Leif Andersson, interview by Andrew Lawler, 2012.

2. The Carnelian Beard

28 *And lo, the Rooster King*: Jay Hopler, "The Rooster King," forthcoming; quoted by permission of the author.

28 *The chicken's grandest entrance*: Daniel T. Potts, *A Companion to the Archaeology of the Ancient Near East* (Chichester, West Sussex: Wiley-Blackwell, 2012), 843; F. S. Bodenheimer, *Animal and Man in Bible Lands* (Leiden: E.J. Brill, 1960), 166; Richard A. Gabriel, *Great Captains of Antiquity* (Westport, CT: Greenwood Press, 2001), 22; Donald B. Redford, *The Wars in Syria and Palestine of Thutmose III* (Boston: Brill, 2003), 225; Nicolas Grimal, *A History of Ancient Egypt* (Paris: Librairie Arthéme Fayard, 1988), 216.

29 *The four exotics*: J. B. Coltherd, "The Domestic Fowl in Ancient Egypt," *Ibis* 108, no. 2 (1966): 217. See also John Dagnell Bury et al. eds., "The Reign of Thutmose III," chapter 4 in *The Cambridge Ancient History: The Egyptian and Hittite Empires to 1000 B.C.*, vol. 2 (New York: Macmillan, 1924).

29 *Only five foot three inches tall*: Homer, "A Visit of Emissaries," book nine in *The Iliad*, trans. Robert Fitzgerald (New York: Farrar, Straus and Giroux, 2004), 209.

29 *But the inscription*: "The reference you have for Thutmose III is tricky since the text of the king's annals at Karnak is damaged. It possibly refers to birds that give birth (ms) every day. The word *ms* is a reconstruction and should be taken with a grain of salt. In any case, regardless of what the text of Thutmose III says, chicken would have been rare exotic animals that may have been kept in personal and/or royal zoos at this time." Rozenn Bailleul-LeSuer, interview by Andrew Lawler, 2013.

29 *They were mesmerized*: Glenn E. Perry, *The History of Egypt* (Westport, CT: Greenwood Press, 2004), 1.

30 *Across the Nile, on the barren*: Prisse D'Avennes and Olaf E. Kaper, *Atlas of Egyptian Art* (Cairo: American University in Cairo Press, 2000), 137.

30 *Working in the Valley of the Kings*: Howard Carter, "An Ostracon Depicting a Red Jungle-Fowl. The Earliest Known Drawing of the Domestic Cock," *The Journal of Egyptian Archaeology* 9, no. 1/2 (1923): 1–4.

30 *Yet the sought-after Egyptologist*: Ibid.

31 *It struts its stuff like any*: Ibid.

31 *Based on its location, he*: Ibid.

31 *The other candidate is a silver*: Christine Lilyquist, "Treasures from Tell Basta," *Metropolitan Museum Journal* 47 (2012): 39.

31 *But if the chicken was special*: Bailleul-LeSuer, interview by Andrew Lawler [9AT2TK].

31 *They also mummified hundreds*: Kathryn A. Bard and Steven Blake Shubert, eds., *Encyclopedia of the Archaeology of Ancient Egypt* (London: Routledge, 1999), s.v., "Saqqara."

32 *While the Egyptians*: Andrew Lawler, "Unmasking the Indus: Boring No More, a Trade-Savvy Indus Emerges," *Science* 320, no. 5881 (June 2008): 1276–1281, doi: 10.1126/science.320.5881.1276.

33 *Richard Meadow and his colleague Ajita Patel*: Richard Meadow and Ajita Patel, interviews by Andrew Lawler, 2013.

34 *West of Delhi*: Vasant Schinde, interview by Andrew Lawler, 2012.

34 *One enterprising archaeologist recently*: Sharri R. Clark, *The Social Lives of Figurines: Recontextualizing the Third Millennium BC Terracotta Figurines from Harappa (Pakistan)*, (Oxford: Oxbow Books, 2012), ch. 4.

35 *But recent analysis of Indus cookware*: Andrew Lawler, "Where Did Curry Come From?" *Slate*, January 29, 2013, accessed March 18, 2014, http://www.slate.com/articles /life/food/2013/01/indus_civilization_food_how_scientists_are_figuring_out_what _curry_was_like.html.

35 *the archaeologist Arunima Kashyap*: Ibid.

36 *At an Indus site called Lothal*: Gregory L. Possehl, *The Indus Civilization: A Contemporary Perspective* (Walnut Creek, CA: AltaMira Press, 2002), 80.

37 *According to Genesis*: Genesis: 11:31 (*New American Bible*).

37 *The city in 2000 BC drew traders*: Daniel T. Potts, *A Companion to the Archaeology of the Ancient Near East*, 707.

37 *Shulgi also is credited with*: Hans Baumann, *In the Land of Ur: The Discovery of Ancient Mesopotamia* (New York: Pantheon Books, 1969), 111.

38 *One tablet, dated to the thirteenth year of Ibbi-Sin's*: Peter Steinkeller, interview by Andrew Lawler, 2013.

38 *A Turk invited to an American Thanksgiving*: Mark Forsyth, "The Turkey's Turkey Connection," *New York Times*, November 27, 2013, accessed March 18, 2014, http:// www.nytimes.com/2013/11/28/opinion/the-turkeys-turkey-connection.html?_r=0.

38 *What is called "spotted" may be "speckled"*: Steinkeller, interview.

38 *Some specialists think that, based on a variety of clues*: Daniel T. Potts, *A Companion to the Archaeology of the Ancient Near East*, 763.

39 *Called "Enki and the World Order"*: J. A. Black et al., "Enki and the World Order: Translation," *The Electronic Text Corpus of Sumerian Literature*, University of Oxford, 1998–, accessed March 18, 2014, http://etcsl.orinst.ox.ac.uk/section1/tr113.htm.

39 *"No, no, the sternum"*: Joris Peters, interview by Andrew Lawler, 2012.

41 *In 1988, for example, a Chinese*: Barbara West and Ben-Xiong Zhou, "Did Chickens Go North? New Evidence for Domestication," abstract, *World's Poultry Science Journal* 45, no. 3 (1989), accessed March 18, 2014, http://journals.cambridge.org/action /displayAbstract?fromPage=online&aid=624516.

41 *The oldest textual evidence for Chinese chickens*: Ian F. Darwin, *Java Cookbook* (Sebastopol, CA: O'Reilly, 2001), 852.

42 *"We are the oldest religion"*: Baba Chawish, interview by Andrew Lawler, 2014.

43 *Scholars say that the religion*: Birgul Acikyildiz, *The Yezidis: The History of a Community, Culture and Religion* (London: I.B. Tauris & Co., 2010), 74.

43 *Tawûsê Melek takes the form*: Ibid.

43 *A massive ziggurat*: Andrew Lawler, "Treasure Under Saddam's Feet," *Discover*, October 2002.

43 *Within one of those graves, beside the skull*: Prudence Oliver Harper, *Assyrian Origins: Discoveries at Ashur on the Tigris; Antiquities in the Vorderasiatisches Museum, Berlin* (New York: Metropolitan Museum of Art, 1995), 84.

43 *In the Old Testament, Assyrians are*: *All Things in the Bible: An Encyclopedia of the Biblical World*, comp. Nancy M. Tischler (Westport, CT: Greenwood Press, 2006), 44.

44 *The little cylindrical box*: Prudence Oliver Harper, *Assyrian Origins: Discoveries at Ashur on the Tigris*, 84.

44 *A temple dedicated to both gods was built around 1500 BC in Assur*: Alvin J. Cottrell, *The Persian Gulf States: A General Survey* (Baltimore: Johns Hopkins University Press, 1980), 422.

44 *Babylon in the sixth century BC*: Irving L. Finkel et al., *Babylon* (Oxford: Oxford University Press, 2009), 11.

44 *The multicolored Etemenanki*: David Asheri et al., *A Commentary on Herodotus Books I–IV* (Oxford: Oxford University Press, 2007), 201.

44 *Marduk had been the patron*: Paul-Alain Beaulieu, "Nabonidus the Mad King: A Reconsideration of His Steles from Harran and Babylon," in *Representations of Political Power: Case Histories from Times of Change and Dissolving Order in the Ancient Near East* (Winona Lake, IN: Eisenbrauns, 2007), 137.

45 *When he returned*: Lisbeth S. Fried, *The Priest and the Great King: Temple-Palace Relations in the Persian Empire* (Winona Lake, IN: Eisenbrauns, 2004), 29.

45 *They granted a measure of self-government to this multiethnic society*: Daniel T. Potts, *The Archaeology of Elam: Formation and Transformation of an Ancient Iranian State* (New York: Cambridge University Press, 1999), 346.

46 *"The cock is created to oppose the demons and sorcerers"*: Richard D. Mann, *The Rise of Mahāsena: The Transformation of Skanda-Karttikeya in North India from the Kusāna to Gupta Empires* (Leiden: Brill, 2012), 127.

46 *The Persians held the rooster in such esteem that it was forbidden*: Maneckji Nusservanji Dhalla, *Zoroastrian Civilization: From the Earliest Times to the Downfall of the Last Zoroastrian Empire, 651 A.D.* (New York: Oxford University Press, 1922), 185.

46 *It banished the sloth-demon*: Ibid.

46 *The bird landed "the death blow"*: Ibid.

46 *There are no contemporary explanations*: Wouter Henkelman, "The Royal Achaemenid Crown," *Archaeologische Mitteilungen aus Iran* 19 (1995/96): 133.

46 *Another Persian sacred*: *The Oxford Encyclopaedia, Or, Dictionary of Arts, Sciences and General Literature*, comps. W. Harris et al. (Bristol: Thoemmes, 2003), s.v. "Costume."

46 *The chicken arrived in Persia*: Donald P. Hansen and Erica Ehrenberg, "The Rooster in Mesopotamia," in *Leaving No Stones Unturned: Essays on the Ancient Near East and Egypt in Honor of Donald P. Hansen* (Winona Lake, IN: Eisenbrauns, 2002), 53.

46 *According to some traditions*: Mary Boyce, *A History of Zoroastrianism* (Leiden: Brill, 1975), 3.

46 *Zoroaster, some scholars say*: Ibid., 192.

46 *Ahura Mazda created Angra Mainyu*: Ibid.

47 *And among Ahura Mazda's assistants*: Touraj Daryaee, *The Oxford Handbook of Iranian History* (Oxford: Oxford University Press, 2012), 91.

47 *One of his tools is the rooster*: Martin Haug and Edward William West, *Essays on the Sacred Language, Writings, and Religion of the Parsis* (London: Trübner & Co., 1878), 245.

47 *Such Zoroastrian beliefs penetrated much*: Andrew Lawler, "Edge of an Empire," *Archaeology Journal* 64, no. 5 (September/October 2011).

47 *A Persian coffin*: Hansen and Ehrenberg, *Leaving No Stones Unturned*, 53.

47 *The Persian prophet's view of*: Risa Levitt Kohn and Rebecca Moore, *A Portable God: The Origin of Judaism and Christianity* (Lanham, MD: Rowman & Littlefield Publishers, 2007), 65.

47 *Before the Persians came to Palestine*: Gunnar G. Sevelius, MD, *The Nine Pillars of History* (Self-published via AuthorHouse, 2012), 237.

47 *The only religious authorities*: Matthew 2:1 (*New American Bible*).

47 *A couple of centuries after Cyrus's*: Simon Davis, *The Archaeology of Animals* (New Haven: Yale University Press, 1987), 187.

47 *The rooster's crow*: Mark 14:72 (*New American Bible*).

47 *By early medieval times*: Accessed May 14, 2014, http://thingstodo.viator.com/vatican -city/st-peters-basilica-sacristy-treasury-museum/.

48 *"When you listen"*: Ram Swarup, *Understanding the Hadith: The Sacred Traditions of Islam* (Amherst, NY: Prometheus Books, 2002), 199.

48 *According to some Islamic traditions*: James Lyman Merrick, *The Life and Religion of Mohammed: As Contained in the Sheeäh Traditions of the Hyât-ul-Kuloob* (Boston: Phillips, Sampson, and Co., 1850), 196.

48 *One scholar thinks that the very shape*: Hansen and Ehrenberg, *Leaving No Stones Unturned*, 61.

48 *The bird's origin in the mysterious*: Maarten Jozef Vermaseren, *The Excavations in the Mithraeum of the Church of Santa Prisca in Rome* (Neiden, Norway: Brill, 1965), 163.

48 *This may reflect Zoroastrian*: Sanping Chen, *Multicultural China in the Early Middle Ages* (Philadelphia: University of Pennsylvania Press, 2012), 110.

48 *A Chinese legend from the*: Roel Sterckx, *The Animal and the Daemon in Early China* (Albany: State University of New York Press, 2002), 42.

48 *Daoist priests in this period*: Ibid., 41.

48 *A jiren, or chicken officer*: Ibid.

49 *When Chinese rulers*: Sanping Chen, *Multicultural China in the Early Middle Ages*, 104.

49 *Even today, one of*: Historical Dictionary of Chinese Cinema, comps. Tan Ye and Yun Zhu (Lanham, MD: Scarecrow Press, 2012), s.v. "Golden Rooster Awards, The."

49 *The Chinese ideogram for rooster*: Deanna Washington, *The Language of Gifts: The Essential Guide to Meaningful Gift Giving* (Berkeley, CA: Conari Press, 2000), 86.

49 *In early medieval Korea*: Han'guk Munhwa Korean Culture (Los Angeles: Korean Cultural Service, 2001), 22:27.

49 *In Japan, by the seventh century AD*: Fukuda Hideko, "From Japan to Ancient Orient," *The Epigraphic Society Occasional Papers*, 1998, 23, 105.

49 *The western Chinese minority group*: Michael Witzel, *The Origins of the World's Mythologies* (Oxford: Oxford University Press, 2012), 144.

49 *A team of Japanese researchers*: Tsuyoshi Shimmura and Takashi Yoshimura, "Circadian Clock Determines the Timing of Rooster Crowing," *Current Biology* 23, no. 6 (2013): R231–233, doi:10.1016/j.cub.2013.02.015.

49 *From Germanic graves*: Leslie Webster and Michelle Brown, *The Transformation of the Roman World AD 400–900* (Berkeley: University of California Press, 1997), 153.

3. The Healing Clutch

50 *"We owe a cock"*: Plato, *Symposium and the Death of Socrates*, trans. Tom Griffith (Ware, Hertfordshire: Wordsworth Edition Limited, 1997), 210.

50 *The last words of a condemned*: Alexander Nehamas, *Virtues of Authenticity: Essays on Plato and Socrates* (Princeton, NJ: Princeton University Press, 1999), 48.

50 *In ancient Greece, sacrificing*: Emma Jeannette Levy Edelstein and Ludwig Edelstein, *Asclepius: Collection and Interpretation of the Testimonies* (Baltimore: Johns Hopkins University Press, 1998), 190.

50 *The German philosopher Friedrich Nietzsche*: Friedrich Wilhelm Nietzsche, *The Pre-Platonic Philosophers*, trans. Greg Whitlock (Urbana: University of Illinois Press, 2001), 261.

50 *Others say he was expressing his*: Friedrich Wilhelm Nietzsche, *Twilight of the Idols, Or, How to Philosophize with a Hammer*, ed. Duncan Large (Oxford: Oxford University Press, 1998), 87.

50 *The classicist Eva Keuls*: Eva C. Keuls, *The Reign of the Phallus: Sexual Politics in Ancient Athens* (New York: Harper & Row, 1985), 79.

51 *"Persian cock!"*: G. Theodoridis, trans., "Aristophanes' 'The Birds,'" Poetry in Translation, 2005, accessed March 18, 2014, http://www.poetryintranslation.com/PITBR /Greek/Birds.htm.

51 *Three centuries earlier, when Homer*: Homer, "A Visit of Emissaries," book nine in *The Iliad*, trans. Robert Fitzgerald (New York: Farrar, Straus and Giroux, 2004), 209.

51 *A detailed and lifelike terra-cotta rooster*: Steven H. Rutledge, *Ancient Rome as a Museum: Power, Identity, and the Culture of Collecting* (Oxford: Oxford University Press, 2012), 88.

51 *Aesop's goose laid a golden egg*: Samuel Croxall and George Fyler Townsend, *The Fables of Aesop* (London: F. Warner, 1866), 61.

51 *"Nourish a cock, but"*: Thomas Taylor, trans., *Iamblichus' Life of Pythagoras, Or, Pythagoric Life* (Rochester, VT: Inner Traditions International, 1986), 207.

51 *The bird also was associated with Persephone*: Mark P. Morford and Robert J. Lenardon, *Classical Mythology* (New York: McKay, 1971), 241.

51 *At the time of Socrates's death*: Pierre Briant, *From Cyrus to Alexander: A History of the Persian Empire* (Winona Lake, IN: Eisenbrauns, 2002), 548.

52 *A deliciously juicy new fruit*: *The Oxford Encyclopedia of Ancient Greece and Rome*, comps. Michael Gagarin and Elaine Fantham (New York: Oxford University Press, 2010), s.v. "Peach."

52 *"It was he"—the rooster—*: Theodoridis, "Aristophanes' 'The Birds.'"

53 *Roosters on Greek vases perch*: Yves Bonnefoy, *Greek and Egyptian Mythologies* (Chicago: University of Chicago Press, 1992), 131.

53 *The helmet of the famous*: J. J. Pollitt, *The Art of Ancient Greece: Sources and Documents* (Cambridge: Cambridge University Press, 1990), 64.

53 *Laying hens were common by Socrates's*: George Moore, *Ancient Greece: A Compre-*

hensive Resource for the Active Study of Ancient Greece (Nuneaton, U.K.: Prim-Ed, 2000), 16.

53 *In the original Hippocratic oath*: *Oxford Encyclopedia of Ancient Greece and Rome*, s.v. "Hippocratic Corpus."

53 *Called alabastrons, these little*: George B. Griffenhagen and Mary Bogard, *History of Drug Containers and Their Labels* (Madison, WI: American Institute of the History of Pharmacy, 1999), 4.

53 *There was a spacious Asklepion*: Sara B. Aleshire, *The Athenian Asklepieion: The People, Their Dedications, and the Inventories* (Amsterdam: J.C. Gieben, 1989).

54 *At Epidaurus*: Helmut Koester, *History, Culture, and Religion of the Hellenistic Age* (Philadelphia: Fortress Press, 1982), 167.

54 *At the Asklepion on the island*: Aleshire, *The Athenian Asklepieion*.

54 *All that remains of this era*: Walter J.Friedlander, *The Golden Wand of Medicine: A History of the Caduceus* (Westport, CT: Greenwood Press, 1992).

54 *But as late as the third century AD*: Thomas Taylor, ed., *Select Works of Porphyry; Containing His Four Books on Abstinence from Animal Food; His Treatise on the Homeric Cave of the Nymphs; and His Auxiliaries to the Perception of Intelligible Natures* (London: T. Rodd, 1823).

54 *Galen, the second-century AD Greek*: Galen, *On the Properties of Food* (Cambridge, Cambridge University Press, 2003), 3.1.10.

54 *Others recommended*: Page Smith and Charles Daniel, *The Chicken Book* (Boston: Little, Brown, 1975), 125.

54 *"Chicken offers so great an advantage to men"*: Ulisse Aldrovandi, *Aldrovandi on Chickens: The Ornithology of Ulisse Aldrovandi 1600*, ed. and trans. L. R. Lind (Norman: University of Oklahoma Press, 1963), 2:259.

55 *In the ancient city of Milas*: Banu Karaoz, "First-aid Home Treatment of Burns Among Children and Some Implications at Milas, Turkey," *Journal of Emergency Nursing* 36, no. 2 (2010): 111–14.

55 *The Roman writer Pliny called it*: Smith and Daniel, *The Chicken Book*, 125.

55 *A 2000 study in* Chest *found*: Bo Rennard et al., "Chicken Soup Inhibits Neutrophil Chemotaxis In Vitro," *Chest* 118, no. 2 (2000): 1150–57.

55 *Another study confirmed that*: K. Saketkhoo et al., "Effects of Drinking Hot Water, Cold Water, and Chicken Soup on Nasal Mucus Velocity and Nasal Airflow Resistance," *Chest* 74 (1978), 74, 408.

55 *And a 2011 study by an*: Matt Kelley, "Top Doctor Says Chicken Soup Does Have Healing Properties," Radio Iowa, January 17, 2011, accessed March 19, 2014, http://www.radio iowa.com/2011/01/17/top-doctor-says-chicken-soup-does-have-healing-properties/.

55 *Rooster combs, for example*: Paul J. Carniol and Neil S. Sadick, *Clinical Procedures in Laser Skin Rejuvenation* (London: Informa Healthcare, 2007), 184.

55 *The pharmaceutical company Pfizer*: Alicia Ault, "From the Head of a Rooster to a Smiling Face Near You," *New York Times*, December 22, 2003, accessed March 19, 2014, http://www.nytimes.com/2003/12/23/science/from-the-head-of-a-rooster-to -a-smiling-face-near-you.html.

55 *And proteins derived from chicken bones*: "Chicken Protein Halts Swelling, Pain of Arthritis Patients in Trial," *Denver Post*, September 24, 1993.

56 *The cargo is from dozens*: Peter Schue, interview by Andrew Lawler, 2013.

56 *Hippocrates described flu symptoms nearly*: Niall Johnson, *Britain and the 1918–19 Influenza Pandemic: A Dark Epilogue* (London: Routledge, 2006), 1; Carol Turking-

ton and Bonnie Ashby, *Encyclopedia of Infectious Diseases* (New York: Facts on File, 1998), 165.

56 *In a normal year, the flu virus sickens millions*: "Influenza (Seasonal)," World Health Organization, March 2014, accessed March 19, 2014, http://www.who.int/mediacentre /factsheets/fs211/en/.

56 *In the global pandemic of 1918*: Niall Johnson, *Britain and the 1918–19 Influenza Pandemic*, 82.

56 *Influenza appears to be a price humans*: Ibid., 33.

56 *Each year, scientists around the world*: "Vaccine Development," Flu.gov, accessed March 19, 2014, http://www.flu.gov/prevention-vaccination/vaccine-development/.

58 *The U.S. Army in World War II*: Richard W. Compans and Walter A. Orenstein, *Vaccines for Pandemic Influenza* (Berlin: Springer, 2009), 49.

59 *American soldiers received*: Ibid., 49.

59 *Until recently, this has been*: "FDA Approves First Seasonal Influenza Vaccine Manufactured Using Cell Culture Technology," U.S. Food and Drug Administration news release, November 20, 2012, accessed March 19, 2014, http://www.fda.gov/news events/newsroom/pressannouncements/ucm328982.htm.

59 *Early the following year, the FDA*: Robert Roos, "FDA Approves First Flu Vaccine Grown in Insect Cells," Center for Infectious Disease Research and Policy, January 17, 2013, accessed March 19, 2014, http://www.cidrap.umn.edu/news-perspective/2013 /01/fda-approves-first-flu-vaccine-grown-insect-cells.

59 *He observed the mating habits of roosters*: Rom Harré, *Great Scientific Experiments: Twenty Experiments That Changed Our View of the World* (Oxford: Oxford University Press, 1983), 31.

60 *Instead, the Greek thinker proposed*: Ibid.

60 *These innovative chicken-egg*: Aristotle, "Book 7: On the Heavens," in *Aristotle's Collection* (Google eBook Publish This, LLC, 2013).

60 *William Harvey in seventeenth-century*: Michael Windelspecht, *Groundbreaking Scientific Experiments, Inventions, and Discoveries of the 19th Century* (Westport, CT: Greenwood Press, 2003), 57.

60 *Marcello Malpighi at Bologna*: Ibid., 167.

60 *Three centuries later, in 1931*: Patrick Collard, *The Development of Microbiology* (Cambridge: Cambridge University Press, 1976), 164.

60 *By the 1950s, cancer researchers*: Andries Zijlstra, interview by Andrew Lawler, 2013.

61 *On October 30, 1878, a package*: Stanley A. Plotkin, ed., *History of Vaccine Development* (New York: Springer, 2010), 35.

61 *"We now have a culture"*: H. Bazin, *Vaccination: A History from Lady Montagu to Genetic Engineering* (Montrouge, France: J. Libbey Eurotext, 2011), 152.

62 *"Chance only favors the"*: Stanley Finger, *Minds Behind the Brain: A History of the Pioneers and Their Discoveries* (Oxford: Oxford University Press, 2000), 309.

62 *By January of 1880, Pasteur*: Plotkin, *History of Vaccine Development*, 36.

62 *"Through certain changes"*: Bazin, *Vaccination*, 163.

62 *A Dutch doctor in Indonesia*: Christiaan Eijkman, *Polyneuritis in Chickens, or the Origin of Vitamin Research* (papers, Hoffman-la Roche, 1990); Henry E. Brady and David Collier, *Rethinking Social Inquiry: Diverse Tools, Shared Standards* (Lanham, MD: Rowman & Littlefield, 2004), 228; Richard Gordon, *The Alarming History of Medicine: Amusing Anecdotes from Hippocrates to Heart Transplants* (New York: St. Martin's Griffin, 1993), 63.

63 *In 1910, the director of New York's*: Harry Bruinius, *Better for All the World: The Secret History of Forced Sterilization and America's Quest for Racial Purity* (New York: Knopf, 2006).

63 *Laughlin saw chicken breeding*: Francis Galton and Karl Pearson, *The Life, Letters and Labours of Francis Galton*, vol. 3, part 1 (Cambridge: Cambridge University Press, 2011), 221.

63 *In 1933, lawmakers in Nazi Germany decreed*: Michael R. Cummings, *Human Heredity: Principles and Issues* (Pacific Grove, CA: Brooks/Cole, 2000), 13.

63 *The French biologist Alexis Carrel*: Andrés Horacio Reggiani, *God's Eugenicist: Alexis Carrel and the Sociobiology of Decline* (New York: Berghahn Books, 2007), 41.

63 *This experiment electrified*: "Living Tissue Endowed by Carrel with 'Eternal Youth' Has Birthday; Begins Today New Year of 'Immortality' in Its Glass 'Olympus' at Laboratory—Age in Human Terms Is Put at 200," *New York Times*, January 16, 1942.

63 *Carrel was a member of a scientific*: H. H. Laughlin, *Report of the Committee to Study and to Report on the Best Practical Means of Cutting Off the Defective Germ-Plasm in the American Population* (Cold Spring Harbor, NY: Eugenics Record Office, 1914).

64 *At Edinburgh's Roslin Institute*: Helen Sang, "*Transgenic Chickens*—Methods and Potential Applications," *Trends in Biotechnology* 12 (1994): 415–20.

4. Essential Gear

65 *But thirty thousand years ago*: Brian M. Fagan and Charlotte Beck, *The Oxford Companion to Archaeology* (New York: Oxford University Press, 1996), 543.

65 *Humans then moved around the Pacific*: Claudia Briones and José Luis Lanata, eds., *Archaeological and Anthropological Perspectives on the Native Peoples of Pampa, Patagonia, and Tierra del Fuego to the Nineteenth Century* (Westport, CT: Bergin & Garvey, 2002), 6.

66 *Not until after AD 1200*: Andrew Lawler, "Beyond *Kon-Tiki*: Did Polynesians Sail to South America?" *Science* 328, no. 5984 (2010): 1344–47, doi: 10.1126/science.328.5984.1344.

66 *"How shall we account for this"*: A. Grenfell Price, ed., *The Explorations of Captain James Cook in the Pacific, as Told by Selections of His Own Journals, 1768–1779* (New York: Dover Publications, 1971), 222.

66 *Well into the twentieth century*: Ben R. Finney, *Voyage of Rediscovery: A Cultural Odyssey through Polynesia* (Berkeley: University of California Press, 1994), 12.

66 *At the insistence of the naturalist Joseph*: Ibid., 7.

66 *"These people sail in those seas"*: Ibid., 11.

66 *"When this comes to be prov'd"*: James Cook, *Captain Cook's Journal During His First Voyage Round the World Made in H.M. Bark "Endeavour" 1768–71*, ed. Captain W. J. L. Whartom (London: Elliot Stock, 1893; Google eBook, 2013).

66 *Aboard the* Endeavour: "HMS Endeavour," Technogypsie.com, April 24, 2011, accessed March 19, 2014, http://www.technogypsie.com/science/?p=200.

66 *For the Polynesians, they*: John Hawkesworth, W. Strahan, and T. Cadell, *An Account of the Voyages Undertaken by the Order of His Present Majesty for Making Discoveries in the Southern Hemisphere*, vol. 2 (London: printed for W. Strahan and T. Cadell in the Strand, 1773; Google eBook).

67 *Easter Islanders made a similar*: Steven R. Fischer, *Island at the End of the World: The Turbulent History of Easter Island* (London: Reaktion, 2005).

67 *"Chickens played an important"*: Scoresby Routledge, *The Mystery of Easter Island: The*

Story of an Expedition (London: printed for the author by Hazell, Watson and Viney, 1919), 218.

67 *By Routledge's day*: Kathy Pelta, *Rediscovering Easter Island* (Minneapolis: Lerner Publications, 2001), 36.

67 *"Humankind's covetousness is"*: Terry Hunt and Carl Lipo, *The Statues That Walked: Unraveling the Mystery of Easter Island* (New York: Free Press, 2011), 20.

67 *Author Jared Diamond*: Jared Diamond, "Easter's End," *Discover*, August 1995.

68 *Hundreds of these structures*: Edwin Ferdon Jr., "Stone Chicken Coops on Easter Island," *Rapa Nui Journal* 14, no. 3 (2000).

68 *In his 1997 book,* Guns: Jared M. Diamond, *Guns, Germs, and Steel: The Fates of Human Societies* (New York: W. W. Norton & Company, 1999), 60.

69 *"No Nation will ever"*: James Cook, *The Journals*, ed. Philip Edwards (London: Penguin 2003; published in Penguin Classics as *The Journals of Captain Cook*, 1999), 337.

69 *He estimated that there were six*: Ibid., 271.

69 *During Cook's visit*: A. Grenfell Price, ed., *The Explorations of Captain James Cook in the Pacific, as Told by Selections of His Own Journals, 1768–1779* (New York: Dover Publications, 1971), 155.

69 *Cook's men also noted*: Ibid.

69 *With the possible exception*: "Elizabeth Taylor Chokes on Bone," *Times-News*, October 13, 1978.

69 *"Why did the chicken"*: John Noble Wilford, "First Chickens in Americas Were Brought from Polynesia," *New York Times*, June 5, 2007, accessed March 19, 2014, http://www.nytimes.com/2007/06/05/science/05chic.html.

70 *The chicken remains*: A. A. Storey et al., "Radiocarbon and DNA Evidence for a Pre-Columbian Introduction of Polynesian Chickens to Chile," *Proceedings of the National Academy of Sciences* 104, no. 25 (2007): 10335–0339, doi:10.1073/pnas.0703993104.

70 *The chicken arrived in Sweden*: Terry L. Jones, *Polynesians in America: Pre-Columbian Contacts with the New World* (Lanham, MD: AltaMira Press, 2011), 142.

70 *The first documented chickens*: Kathleen A. Deagan and José María Cruxent, *Archaeology at La Isabela: America's First European Town* (New Haven: Yale University Press, 2002), 5.

71 *The Spanish conquistador Hernán Cortés*: Jones, *Polynesians in America*, 144.

71 *That zoo, built of marble*: *Pacific Discovery*, 7–9 (California Academy of Sciences, 1955): 164.

71 *Cortés noted a street*: Hernán Cortés, *Letters of Cortés: The Five Letters of Relation from Fernando Cortes to the Emperor Charles V*, ed. and trans. Francis Augustus MacNutt (Cleveland: Arthur H. Clark, 1908), 257.

71 *The Nahuatl-speaking*: James Lockheart, *The Nahuas after the Conquest: A Social and Cultural History of the Indians of Central Mexico, Sixteenth through Eighteenth Centuries* (Stanford, CA: Stanford University Press, 1992), 278.

71 *When a Portuguese navigator*: Jones, *Polynesians in America*, 160.

71 *Two decades later*: Antonio Pigafetta, *The First Voyage around the World (1519–1522): An Account of Magellan's Expedition*, ed. Theodore J. Cachey (New York: Marsilio Publishers, 1995), 8.

71 *In 1527, a Spanish*: Alida C. Metcalf, *Go-betweens and the Colonization of Brazil: 1500–1600* (Austin: University of Texas Press, 2005), 127.

72 *In 1848, as the biologist*: Alfred Russell Wallace, *Travels on the Amazon and Rio Negro:*

With an Account of the Native Tribes and Observations on the Climate, Geology, and Natural History of the Amazon Valley (London: Ward Lock, 1889), 210.

72 *One intriguing report*: Jones, *Polynesians in America*, 145.

72 *Just two years earlier*: Domingo Martinez-Castilla to University of Missouri–Columbia colleagues, December 15, 1996, accessed March 19, 2014, http://www.andes.missouri .edu/Personal/DMartinez/Diffusion/msg00028.html.

72 *A half century later*: Raul Borras Barrenechea, ed. *Relacion del Descubrimiento del Reyno del Peru* (Lima: Instituto Raul Porras Barrenechea 1970), 41–60.

72 *But the very name*: Jones, *Polynesians in America*, 52.

72 *A museum in Lima*: "Zoomorphic Polychrome Terracotta Vessel in Shape of Rooster, Peru, Vicus Culture, Pre–Inca Civilization, circa 100 B.C.," The Bridgeman Art Library, accessed March 19, 2014, http://www.bridgemanart.com/en-GB/asset/512719.

73 *One Jesuit in the 1580s*: Metcalf, *Go-betweens and the Colonization of Brazil*,152.

73 *By the close of the sixteenth*: W. S. W. Ruschenberger, *Three Years in the Pacific; Including Notices of Brazil, Chile, Bolivia, Peru* (Philadelphia: Carey, Lea & Blanchard, 1834), 394.

73 *Bantu slaves from southern*: Jones, *Polynesians in America*, 145.

73 *Those bones had*: Lisa Matisoo-Smith, email message to author, 2013.

74 *"An ancient Polynesian haplotype"*: Storey, "Radiocarbon and DNA Evidence."

74 *When we eat and*: Sheridan Bowman, *Radiocarbon Dating* (Berkeley: University of California Press, 1990), 25.

75 *Gongora and others*: Gongora et al., "Indo-European and Asian Origins for Chilean and Pacific Chickens Revealed by MtDNA," *Proceedings of the National Academy of Sciences of the United States of America* 105, no. 30 (2008): 10308, doi:10.1073 /pnas.0807512105.

75 *Matisoo-Smith, Storey*: Alice A. Storey et al., "Pre-Columbian Chickens of the Americas: A Critical Review of the Hypotheses and Evidence for Their Origins," *Rapa Nui Journal* 25 (2011): 5–19.

75 *An independent team of molecular biologists*: Scott M. Fitzpatrick and Richard Callaghan, "Examining Dispersal Mechanisms for the Translocation of Chicken (*Gallus Gallus*) from Polynesia to South America," *Journal of Archaeological Science* 36, no. 2 (2009): 214–23, doi:10.1016/j.jas.2008.09.002.

75 *Recent genetic work*: Caroline Roullier et al., "Historical Collections Reveal Patterns of Diffusion of Sweet Potato in Oceania Obscured by Modern Plant Movements and Recombination," *Proceedings of the National Academy of Sciences of the United States of America* 110, no. 6 (2013): 2205–210, doi:10.1073/pnas.1211049110.

76 *Wind and current studies*: Jones, *Polynesians in America*, 173.

76 *Historical records mention*: Finney, *Voyage of Rediscovery*, 1994.

76 *Words, tools, ritual objects*: A. Lawler, "Northern Exposure in Doubt," *Science* 328, no. 5984 (2010): 1347.

76 *Island myths, many centered*: W. D. Westervelt, *Legends of Old Honolulu: Collected and Translated from the Hawaiian* (London: Constable & Co., 1915), 230.

76 *Archaeologists working*: Jones, *Polynesians in America*, 125.

77 *Looming in the distance*: *Maha'ulepu, Island of Kaua'i Reconnaissance Survey*, U.S. Department of the Interior, National Park Service, Pacific West Region, Honolulu Office, February 2008; Edward Tregear, " 'The Creation Song' of Hawaii," *The Journal of the Polynesian Society* 9, no. 1 (March 1900): 38–46.

77 *The archaeologist David Burney*: David Burney, interview by Andrew Lawler, 2013.

78 *Now researchers are pulling DNA*: Nicholas Wade, "Dead for 32,000 Years, an Arctic Plant Is Revived," *New York Times*, February 20, 2012, accessed March 19, 2014, http://www.nytimes.com/2012/02/21/science/new-life-from-an-arctic-flower-that -died-32000-years-ago.html.

80 *"You end up with"*: Peggy Macqueen, interview by Andrew Lawler, 2013.

81 *Cockfighting was associated*: William Ellis, *Polynesian Researches during a Residence of Nearly Eight Years in the Society and Sandwich Islands* (London: Bohn, 1853), 223.

82 *"The Pacific is a basket"*: Alan Cooper, interview by Andrew Lawler, 2013.

82 *Cooper's team extracted DNA*: Ibid.

83 *By 1200 BC, when Ramses*: Paul Wallin and Helene Martinsson-Wallin, eds., *The Gotland Papers: Selected Papers from the VII International Conference on Easter Island and the Pacific: Migration, Identity, and Cultural Heritage* (Gotland University, Sweden: Gotland University Press 11, 2007), 210.

83 *There they remained until Samoa and Tonga*: Jared M. Diamond, *Natural Experiments of History* (Cambridge, MA: Belknap Press of Harvard University Press, 2011), 48.

83 *The last burst of movement*: Neil Asher Silberman and Alexander A. Bauer, *The Oxford Companion to Archaeology* (New York: Oxford University Press, 2012), 660.

84 *Archaeologists call it Lapita*: Ibid., 210.

84 *One view is that the Lapita*: Ibid., 592.

84 *Cooper's team has intriguing*: Vicki Thomson et al., "Using Ancient DNA to Study the Origins and Dispersal of Ancestral Polynesian Chickens across the Pacific," *Proceedings of the National Academy of Sciences of the United States of America*, March 24, 2014, 113, no. 13 (2014): 4826.

5. Thrilla in Manila

86 *The World Slasher Cup*: Rolando Luzong, interview by Andrew Lawler, 2013.

91 *Filipinos raise "large cocks"*: Alfredo R. Roces, *Filipino Heritage: The Making of a Nation* (Manila: Lahing Pilipino Pub., 1978), 1591.

91 *When the hungry and*: Antonio Pigafetta, *Magellan's Voyage: A Narrative Account of the First Circumnavigation*, ed. and trans. Raleigh Ashlin Skelton (New York: Dover Publications, 1994), 65.

91 *Pigafetta was one of*: Donald F. Lach, *The Century of Discovery of Asia in the Making of Europe*, vol. 1 (Chicago: University of Chicago Press, 1994), 639.

92 *The first three attempts at colonization*: Luis Francia, *A History of the Philippines: From Indios Bravos to Filipinos* (New York: Overlook Press, 2010), 55.

92 *That lucrative trade: Southeast Asia*: Ooi Keat Gin, ed., *A Historical Encyclopedia, from Angkor Wat to East Timor* (Santa Barbara, CA: ABC-CLIO, 2004), s.v. "Spanish Philippines."

92 *Scattered populations were*: Francia, *A History of the Philippines*, 64.

92 *"The sight is one"*: Fedor Jagor, *Travels in the Philippines* (London: Chapman and Hall, 1875), 28.

93 *According to some historians*: Ricky Nations, "The 'Gypsy Chickens' of Key West," *The Southernmost Point* (blog), October 14, 2013, accessed March 19, 2014, http://nations southernmostpoint.blogspot.com/2013/10/the-gypsy-chickens-of-key-west.html.

93 *As early as the 1700s*: Attorney-General, ed., *Official Opinions of the Attorney-General of the Philippine Islands Advising the Civil Governor, the Heads of Departments, and Other Public Officials in Relation to Their Official Duties* (Manila: Bureau of Public Printing, 1903), 638.

93 *Cockpit licenses combined with game-fowl*: Charles Burke Elliott, *The Philippines to the End of the Military Regime* (Indianapolis: Bobbs-Merrill Company, 1916), 263.

93 *In 1861, Madrid's revenues*: Ibid., 263.

93 *"It being a vice, the laws permit"*: Frank Charles Laubach, *The People of the Philippines, Their Religious Progress and Preparation for Spiritual Leadership in the Far East* (New York: George H. Doran, 1925), 403.

93 *"To the cockpit went"*: José Rizal, *Noli me tangere (Touch Me Not)*, ed. and trans. Harold Augenbraum (New York: Penguin, 2006), 302.

94 *Cockfighting is forbidden*: "Republic Act No. 229," *Official Gazette* 44, no. 8, August 1948, accessed March 19, 2014, http://www.gov.ph/1948/06/09/republic-act-no-229/.

94 *"How can a people"*: Alan Dundes, *The Cockfight: A Casebook* (Madison: University of Wisconsin Press, 1994), 139.

94 *Kincaid was an American lawyer*: *Report of the Philippine Commission to the Secretary of War: 1910* (Washington, D.C.: Government Printing Office, 1911), 421.

94 *"Cockfighting has spread to such"*: Ibid., 415.

94 *"Baseball is a tremendous"*: Edward Thomas Devine, ed., *The Survey* 37 (October 7, 1916): 19.

95 *When the American author Wallace Stegner*: Wallace Stegner, *Collected Stories* (New York: Penguin, 2006), 372.

95 *Fearful of large gatherings*: "Philippine Law: Cockfighting Law of 1974," *Gameness til the End* (blog), accessed March 19, 2014, http://gtte.wordpress.com/2011/06/19/philippine-law-cockfighting-law-of-1974/.

95 *His wife, Imelda*: Luzong, interview.

95 *Then, in 1997*: Ibid.

96 *The island of Macao*: Terri C. Walker, *The 2000 Casino and Gaming Business Market Research Handbook* (Norcross, GA: Richard K. Miller and Associates, 2000), 352.

96 *Some sources say that*: *Philippines Free Press* 62, no. 14–26 (1969), 68.

97 *Animal-rights activists claim*: Victoria Maranan, "Gamefowl Breeders Convention Ruffles Feathers," *KXII*, August 11, 2011, accessed March 24, 2014, http://www.kxii.com/news/headlines/Humane_society_accuse_gamefowl_breeders_association_for_illegal_activity_127567283.html; *Animal Fighting Prohibition Enforcement Act of 2005: Hearing on H.R. 817, May 18, 2006, Before the Subcommittee on Crime, Terrorism, and Homeland Security of the Committee on the Judiciary*, 109th Cong., (2006), 20.

97 *Police arrested game-fowl*: Ngoc Nguyen, "Ind. Man Arrested After Story in Filipino Cockfight Magazine," New America Media, August 10, 2010, accessed March 19, 2014, http://newamericamedia.org/2010/08/ind-man-arrested-after-story-in-filipino-cockfight-magazine.php.

98 *Some 15 million game fowl*: Rolando Luzong, "Bantay-Sabong Special Report," Bantay-Sabong's Facebook page, July 14, 2012, accessed March 19, 2014, https://www.facebook.com/permalink.php?id=398215130241725&story_fbid=494144780601019.

98 *The flamboyant buttercup*: "History of Breeds," University of Illinois Extension, Incubation and Embryology, accessed March 19, 2014, http://urbanext.illinois.edu/eggs/res10-breedhistory.html.

99 *The research provided*: Freyja Imsland et al., "The Rose-comb Mutation in Chickens Constitutes a Structural Rearrangement Causing Both Altered Comb Morphology and Defective Sperm Motility," *PLoS Genetics* 8, no. 6 (2012): E1002775, doi:10.1371/journal.pgen.1002775.

99 *In northern Thailand, for*: "Trance Dancing and Spirit Possession in Northern Thailand," *Sanuk* (blog), November 19, 2010, accessed March 19, 2014, http://sanuksanuk .wordpress.com/2010/11/19/trance-dancing-and-spirit-possession-in-northern -thailand/.

100 *One of the earliest recorded*: Robert Joe Cutter, *The Brush and the Spur: Chinese Culture and the Cockfight* (Hong Kong: Chinese University Press, 1989), 10.

100 *"Thus from which"*: Ibid., 14.

100 *In a tomb just outside*: J. Maxwell Miller and John H. Hayes, *A History of Ancient Israel and Judah* (Philadelphia: Westminster Press, 1986), 422.

100 *Another seal with a fighting*: K. A. D. Smelik, *Writings from Ancient Israel: A Handbook of Historical and Religious Documents* (Louisville, KY: Westminster/John Knox Press, 1991), 140.

100 *Philistines in the nearby port*: Paula Hesse, interview by Andrew Lawler, 2013.

100 *A Chinese Daoist*: Louis Komjathy, "Works Consulted and Further Reading," in "Animals and Daoism," *Advocacy for Animals* (blog), September 26, 2011, accessed March 20, 2014, http://advocacy.britannica.com/blog/advocacy/2011/09/daoism -and-animals/.

100 *At the same time, in ancient Greece*: Judith M. Barringer, *The Hunt in Ancient Greece* (Baltimore: Johns Hopkins University Press, 2001), 90.

101 *The birds adorned the high*: Fredrick J. Simons, *Eat Not This Flesh: Food Avoidances from Prehistory to the Present* (Madison: University of Wisconsin Press, 1994), 154.

101 *In India, British officers*: Linda Colley, *Captives: Britain, Empire and the World, 1600–1850* (London: J. Cape, 2002), 349.

101 *An English visitor to China*: R. P. Forster, *Collection of the Most Celebrated Voyages and Travels from the Discovery of America to the Present Time*, vol. 3 (Google eBook: 1818), 321.

101 *A European visitor to early*: Eric Dunning, *Sport Matters: Sociological Studies of Sport, Violence and Civilisation* (London: Routledge, 1999).

101 *King Henry VIII built*: Sarah Stanton and Martin Banham, *Cambridge Paperback Guide to Theatre* (Cambridge: Cambridge University Press, 1996), 72.

102 *James I was a cocker*: Joseph Strutt and William Hone, *The Sports and Pastimes of the People of England: Including the Rural and Domestic Recreations, May Games, Mummeries, Shows, Processions, Pageants, and Pompous Spectacles, from the Earliest Period to the Present Time* (London: printed for Thomas Tegg, 1841), 282.

102 *"Can this cock-pit"*: Albert Rolls, *Henry V* (New York: Infobase Publishing, 2010), 251.

102 *The Globe Theatre*: "Entertainment at Shakespeare's Globe Theatre," No Sweat Shakespeare, accessed March 20, 2014, http://www.nosweatshakespeare.com/resources /globe-theatre-entertainment/; William Shakespeare, *The Yale Shakespeare*, Wilbur L. Cross and Tucker Brooke, eds. (New Haven: Yale University Press, 1918), 122.

102 *The diarist Samuel Pepys*: Samuel Pepys, *The Diary of Samuel Pepys* (New York: Croscup & Sterling, 1900), 385.

102 *"It is wonderful to see the courage"*: Edward Walford, *Old and New London: A Narrative of Its History, Its People, and Its Places* (London: Cassell, 1879), 375.

102 *A Scottish writer in*: William Edward Hartpole Lecky, *A History of England in the Eighteenth Century*, vol. 1 (London: Longmans, Green, and Co. 1878), Online Library of Liberty, accessed March 20, 2014, http://oll.libertyfund.org/?option=com_staticx t&staticfile=show.php%3Ftitle=2035&chapter=145242&layout=html.

102 *At Newcastle upon Tyne*: Tony Collins et al., eds., *Encyclopedia of Traditional British Rural Sports* (London: Routledge, 2005), s.v. "Cockfighting."

102 *William Hogarth's 1759*: Frederic George Stephens and M. Dorothy George, eds., *Catalogue of Prints and Drawings in the British Museum* (London: By Order of the Trustees, 1870), 1223.

102 *When Parliament banned*: "Police Magistrates, Metropolis Act 1833," Animal Rights History, accessed March 20, 2014, http://www.animalrightshistory.org/animal -rights-law/romantic-legislation/1833-uk-act-police-metropolis.htm.

103 *"The old story about"*: Robert Boddice, interview by Andrew Lawler, 2013; see Rob Boddice, *A History of Attitudes and Behaviours toward Animals in Eighteenth- and Nineteenth-century Britain: Anthropocentrism and the Emergence of Animals* (Lewiston, NY: Edwin Mellen Press, 2008).

103 *The British Parliament*: Boddice, *A History of Attitudes and Behaviors*, 22.

103 *Though cockfighting*: "Hunting Act 2004," The National Archives, accessed March 20, 2014, http://www.legislation.gov.uk/ukpga/2004/37/contents.

103 *After dining in Williamsburg*: George Washington, *The Daily Journal of Major George Washington, in 1751–2*, ed. Joseph M. Toner (Albany, NY: J. Munsell's Sons, 1892), 76.

103 *That same year, the capital's*: Ed Crews, "Once Popular and Socially Acceptable: Cock-fighting," *Colonial Williamsburg*, Autumn 2008, accessed March 20, 2014, http:// www.history.org/Foundation/journal/Autumn08/rooster.cfm.

103 *The Virginia General Assembly*: Gerald R. Gems et al., *Sports in American History: From Colonization to Globalization* (Champaign, IL: Human Kinetics, 2008), 1; *Proceedings of the First Provincial Congress of Georgia, 1775: Proceedings of the Georgia Council of Safety, 1775 to 1777; Account of the Siege of Savannah, 1779, from a British Source* (Savannah, GA: Savannah Chapter of the Daughters of the American Revolution, 1901), 7.

104 *In 1782, a twenty-eight-year-old*: "Third Great Seal Committee—May 1782," accessed March 20, 2014, http://www.greatseal.com/committees/thirdcomm/.

104 *Thomas Jefferson avoided*: *Encyclopedia Virginia*, s.v. " 'Life of Isaac Jefferson of Petersburg, Virginia, Blacksmith' by Isaac Jefferson (1847)," last modified May 3, 2013, accessed March 20, 2014, http://www.encyclopediavirginia.org/_Life_of_Isaac _Jefferson_of_Petersburg_Virginia_Blacksmith_by_Isaac_Jefferson_1847; Fawn McKay Brodie, *Thomas Jefferson, an Intimate History* (New York: W. W. Norton, 1974), 63.

104 *"His passions are terrible"*: H. W. Brands, *Andrew Jackson: His Life and Times* (New York: Doubleday, 2005), 97.

104 *He is quoted as saying*: T. F. Schwartz, *For the People: A Newsletter of the Abraham Lincoln Association*, Springfield, IL, Spring 2003, 5:1.

104 *Mark Twain watched a match*: Mark Twain and Charles Dudley Warner, *The Writings of Mark Twain* (New York: Harper & Bros., 1915), 340.

104 *The media magnate William Randolph Hearst*: William Randolph Hearst, *William Randolph Hearst, a Portrait in His Own Words*, ed. Edmond D. Coblentz (New York: Simon & Schuster, 1952), 239.

104 *Not until 2008*: Ed Anderson, "Louisiana's Ban on Cockfighting Takes Effect Friday," *The Times-Picayune*, August 12, 2008, last modified October 12, 2009, accessed March 20, 2014, http://www.nola.com/news/index.ssf/2008/08/louisianas_ban_on _cockfighting.html.

105 *Two large cockpit*: "The Newport Plain Talk—Print Story," *The Newport Plain Talk*, accessed March 20, 2014, http://newportplaintalk.com/printstory/10546.

105 *Thomas Farrow, who led*: J. J. Stambaugh, "Strategy, Stealth Key for FBI in Cocke County Investigative Work," *Knoxville News Sentinel*, October 5, 2008, accessed March 20, 2014, http://www.knoxnews.com/news/2008/Oct/05/a-tough-case-to -crack/.

105 *In 2013, a Tennessee*: Hank Hayes, "Subcommittee Kills Bill to Raise Cockfighting Fine in Tennessee," *Kingsport Times-News*, April 14, 2011, accessed March 20, 2014, http://www.timesnews.net/article/9031289/subcommittee-kills-bill-to-raise -cockfighting-fine-in-tennessee.

105 *"Slavery also is a"*: Jon Lundberg, interview by Andrew Lawler, 2013.

105 *More than a century and a half*: Sam Youngman and Janet Patton, "Cockfighting Enthusiasts Angry with McConnell for Supporting Farm Bill That Stiffens Penalties," *Lexington Herald Leader*, February 19, 2014.

105 *Bevin attended a pro-cockfighting rally*: John Boel, "Politicians at Cockfighting Rally Caught on Video," April 24, 2014, last modified June 8, 2014, accessed May 15, 2014, http://www.wave3.com/story/25336346/politicians-not-chicken-to-support-the -right-to-cockfight#.U1nZ1fxJI.qA.twitter.

105 *His attendance and comments*: Page One, "Everything About Bevin Is a Giant Contradiction," April 29, 2014, accessed May 15, 2014, http://pageonekentucky .com/2014/04/29/everything-about-bevin-is-a-giant-contradiction/?utm_source= feedburner&utm_medium=feed&utm_campaign=Feed%3A+PageOne+(Page+One).

106 *But today, high-tech drugs*: Congressional Record, V. 153, PT. 6, March 26, 2007, to April 17, 2007, 7644.

106 *On the outskirts*: Lorenzo Fragiel, interview by Andrew Lawler, 2013.

6. Giants upon the Scene

109 *This blessed day will I go*: Herman Melville, "Cock-a-Doodle-Doo!" *Harper's Magazine* 8 (1854): 80.

109 *Few Westerners took Marco Polo*: Stephen G. Haw, *Marco Polo's China: A Venetian in the Realm of Khubilai Khan* (London: Routledge, 2006), 130.

110 *"Perhaps no officer of equal"*: *Dictionary of National Biography*, eds. Leslie Stephen and Sidney Lee (London: Smith, Elder and Co., 1885), s.v. "Belcher, Sir Edward (1799–1877)."

110 *The Dictionary*: Basil Stuart-Stubbs, "Belcher, Sir Edward," in *Dictionary of Canadian Biography*, vol. 10, accessed March 20, 2014, http://www.biographi.ca/en/bio /belcher_edward_10E.html.

110 *Belcher was from*: Anonymous, *Men of the Time: Biographical Sketches of Eminent Living Characters . . . Also Biographical Sketches of Celebrated Women of the Time* (London: Kent, 1859), 55.

110 *Shortly after, Diana Belcher*: Edward Belcher et al., *A Report of the Judgment: Delivered on the Sixth Day of June, 1835* (London: Saunders and Benning, 1835).

110 *Diana ultimately got her*: Diana Jolliffe Belcher, *The Mutineers of the Bounty and Their Descendants in Pitcairn and Norfolk Islands* (New York: Harper & Bros., 1871).

111 *His luck changed in 1836*: Richard Brinsley Hinds et al., *The Zoology of the Voyage of H.M.S. Sulphur: Under the Command of Captain Sir Edward Belcher during the Years 1836–42* (London: Smith, Elder, 1844), 2.

111 *"He was much attached"*: Ibid., 2.

111 *A British zoologist named*: Bo Beolens et al., *The Eponym Dictionary of Reptiles* (Baltimore: Johns Hopkins University Press, 2011), s.v. "Belcher."

111 *It is the world's*: Cindy Blobaum, *Awesome Snake Science: 40 Activities for Learning about Snakes* (Chicago: Chicago Review Press, 2012), 84.

111 *Chinese authorities insisted*: Andrew L. Cherry et al., *Substance Abuse: A Global View* (Westport, CT: Greenwood Press, 2002), 41.

112 *In January 1841*: Edward Belcher, *Narrative of a Voyage Round the World* (London: H. Colburn, 1843), 139.

112 *After an absence of six years*: L. S. Dawson, *Memoirs of Hydrography, Including Brief Biographies of the Principal Officers Who Have Served in H.M. Naval Surveying Service between the Years 1750 and 1885* (Eastbourne: Henry W. Keay, the "Imperial Library," 1885; Google eBook), 18.

112 *She enjoyed what she called*: R. J. Hoage and William A. Deiss, *New Worlds, New Animals: From Menagerie to Zoological Park in the Nineteenth Century* (Baltimore: Johns Hopkins University Press, 1996), 50.

112 *"The Orang Outang is too"*: The Royal Archives, *Queen Victoria's Journals*, May 27, 1842, accessed May 18, 2014, www.queenvictoriajournals.org.

112 *Darwin, who had also visited the same ape*: Steve Jones, *The Darwin Archipelago: The Naturalist's Career Beyond Origin of the Species* (New Haven: Yale University Press, 2011), 1.

113 *The queen loved the circus*: S. L. Kotar and J. E. Gessler, *The Rise of the American Circus, 1716–1899* (Jefferson, NC: McFarland & Co., 2011), 132.

113 *The year before*: "The Court," *The Spectator* 16 (London: F.C. Westley, 1843): 50.

113 *In late September 1842*: Sidney Lee, *Queen Victoria* (New York: Macmillan, 1903), 139–44.

113 *The five hens and two cocks*: There is no direct evidence of formal presentation of the birds, and there is continued controversy over when the fowl arrived at Windsor, their breed, and their donor. Several contemporary sources, however, support the assertion that Captain Belcher, newly arrived in London, made the presentation, though it is unlikely this was done in person given his personal circumstances. No other theory can account for the timing of the gift or the peculiar silence about the donor's identity.

113 *The birds were quickly dubbed*: *Illustrated London News* 3–4, December 23, 1843, 409; *The Countryman* 69, no. 2 (1968), 350.

114 *He had noted in his log the purchase of chickens*: Belcher, *Narrative of a Voyage Round the World*, 257.

114 *Victoria and Albert immediately*: Jane Roberts, *Royal Landscape: The Gardens and Parks of Windsor* (New Haven: Yale University Press, 1997), 205.

114 *Raised on a German country estate with*: Ibid., 205.

114 *In December, the government*: Ibid.

114 *"Partly forester, partly builder, partly farmer, and partly gardener"*: Robert Rhodes James, *Prince Albert: A Biography* (New York: Knopf, 1984), 142.

115 *"We walked down to the Farm"*: *Queen Victoria's Journals*, January 23, 1843.

115 *The fanciful structure was*: W. C. L. Martin, *The Poultry Yard: Comprising the Management of All Kinds of Fowls* (London: Routledge, 1852), 7.

115 *When the* Illustrated London News: *Illustrated London News* 3–4, December 23, 1843, 409.

115 *The morning the article was published*: *Queen Victoria's Journals*, December 23, 1843.

115 *The same week*: Charles Dickens, *A Christmas Carol in Prose: Being a Ghost Story of Christmas* (London: Chapman & Hall, 1843), i.

115 *But it was chicken rather*: "A Royal Banquet," *Carlisle Patriot*, January 6, 1844.

116 *The following spring, after*: *Queen Victoria's Journals*, April 4, 1844.

116 *By the summer of 1844*: Ibid., July 12, 1844.

116 *Mainly, however, this*: Ibid., November 22, 1847.

116 *The author of the 1844*: John French Burke, *Farming for Ladies; Or, a Guide to the Poultry-Yard, the Dairy and Piggery* (London: John J. Murray, 1844).

116 *Chickens were already in Britain*: Kitty Chisholm and John Ferguson, *Rome* (Oxford: Oxford University Press in Association with the Open University Press, 1981), 595; Margaret Visser, *Much Depends on Dinner: The Extraordinary History and Mythology, Allure and Obsessions, Perils and Taboos, of an Ordinary Meal* (New York: Grove Press, 1987), 123.

117 *Roman men carried the right*: Janet Vorwald Dohner, *The Encyclopedia of Historic and Endangered Livestock and Poultry Breeds* (Yale University Press, 2001), s.v. "Chickens."

117 *The oldest handwritten*: C. R. Whittaker, *Rome and Its Frontiers: The Dynamics of Empire* (London: Routledge, 2004), 98.

117 *The Latin saying*: John G. Robertson, *Robertson's Words for a Modern Age: A Cross Reference of Latin and Greek Combining Elements* (Eugene, OR: Senior Scribe Publications, 1991), 237.

117 *and Roman bakers*: Apicius, *Cookery and Dining in Imperial Rome*, ed. and trans. Joseph Dommers Vehling (Milton Keynes, U.K.: Lightning Source, 2009), 95.

117 *And in Roman Britain*: Bruce Watson and N. C. W. Bateman, *Roman London: Recent Archaeological Work; Including Papers Given at a Seminar Held at the Museum of England on 16 November, 1996* (Portsmouth, RI: Journal of Roman Archaeology, 1998), 96.

117 *The sixth-century AD Rule*: Terrence Kardong, *Benedict's Rule: A Translation and Commentary* (Collegeville, MN: Liturgical Press, 1996), 326.

117 *It was, the food scholar*: C. Anne Wilson, *Food & Drink in Britain: From the Stone Age to the 19th Century* (Chicago: Academy Chicago Publishers, 1991), 130.

117 *You could buy a whole bird*: Ibid.

118 *In thirteenth-century London*: Ibid., 123.

118 *One historian calculates that*: Phillip Slavin, "Chicken Husbandry in Late-Medieval Eastern England: 1250–1400," *Anthropozoologica* 44, no. 2 (2009): 35–56, doi:10.5252/az2009n2a2.

118 *"Whoever could afford, substituted chickens"*: Ibid.

118 *At Henry VI's 1429*: John Lawrence et al., *Moubray's Treatise on Domestic and Ornamental Poultry: A Practical Guide to the History, Breeding, Rearing, Feeding, Fattening, and General Management of Fowls and Pigeons* (London: Arthur Hall, Virtue, and Co., 1854), 27.

118 *Besides providing delicious meat*: Jeffery L. Forgeng, *Daily Life in Elizabethan England* (Westport, CT: Greenwood Press, 1995), 113.

118 *Their profits, sniffed*: Samuel Smith, *General View of the Agriculture of Galloway Comprehending Two Counties, Viz. the Stewartry of Kirkcudbright and Wigtonshire, with Observations on the Means of Their Improvement* (London: printed for Sherwood, Neely, and Jones, 1813), 298.

118 *In 1801, Parliament*: David W. Galenson, *Markets in History: Economic Studies of the Past* (Cambridge: Cambridge University Press, 1989), 16.

118 *In 1825, London's population*: Jennifer Speake, *Literature of Travel and Exploration: An Encyclopedia* (New York: Fitzroy Dearborn, 2003), 739.

119 *Less food and more*: Thomas Robert Malthus and Michael P. Fogarty, *An Essay on the Principle of Population: In Two Volumes* (London: Dent, 1967), 15.

119 *One of his tutors, Adolphe Quetelet*: Gillian Gill, *We Two: Victoria and Albert: Rulers, Partners, Rivals* (New York: Ballantine Books, 2009), 134.

119 *The prince encouraged*: Roberts, *Royal Landscape*, 93.

119 *The transformation of British*: J. W. Reginald Hammond, *Complete England* (London: Ward Lock, 1974), 20.

120 *By the 1840s Leadenhall*: George Dodd, *The Food of London: A Sketch of the Chief Varieties, Sources of Supply, Probable Quantities, Modes of Arrival, Processes of Manufacture, Suspected Adulteration, and Machinery of Distribution, of the Food for a Community of Two Millions and a Half* (London: Longmans, Brown, Green and Longmans, 1856), 326.

120 *Most poultry was brought*: Ibid.

120 *Britons ate about 60 million*: William Henry Chandler, *Chandler's Encyclopaedia: An Epitome of Universal Knowledge* (New York: Collier, 1898), vol. 5; s.v. "Poultry."

120 *Eggs were also used to*: Lawrence, *Moubray's Treatise on Domestic and Ornamental Poultry*, 48.

120 *As the Windsor aviary*: Illustrated.

121 *"In order to improve"*: Berkshire Chronicle, September 28, 1844.

121 *London's first poultry show*: Poultry Science 47, 1968, 1–1048.

121 *Europe's wet June continued*: Charles C. Mann, *1493: Uncovering the New World Columbus Created* (New York: Knopf, 2011), 285.

121 *"Another fine morning, when"*: Queen Victoria's Journals, September 13, 1845.

122 *That day, the potato blight*: Mann, *1493*, 285.

122 *An 1845 government report*: Margaret F. Sullivan, *Ireland of To-day; the Causes and Aims of Irish Agitation* (Philadelphia: J.C. McCurdy & Co., 1881), 185.

122 *"Deprive him of this"*: Joseph Fisher, *The History of Landholding in Ireland* (London: Longmans, Green, 1877), 119.

122 *At Windsor on November 6*: Queen Victoria's Journals, November 6, 1845.

122 *"Half of the potatoes"*: James H. Murphy, *Abject Loyalty: Nationalism and Monarchy in Ireland During the Reign of Queen Victoria* (Washington, DC: Catholic University of America Press, 2001), 62.

122 *That winter, outraged Irish*: Michael Gillespie, *The Theoretical Solution to the British/Irish Problem: Using the General Theory of a Federal Kingdom Clearly Stated and Fully Discussed in This Thesis* (Bloomington, IN: AuthorHouse, 2013), 115.

123 *In 1841, the Irish sent*: Fisher, *History of Landholding*, 118.

123 *"Eggs also constitute"*: Lawrence, *Moubray's Treatise on Domestic and Ornamental Poultry*, 48.

123 *On February 23, 1846*: John Kelly, *The Graves Are Walking: The Great Famine and the Saga of the Irish People* (New York: Henry Holt, 2012), 75.

123 *The Royal Dublin Society*: The Journal of the Royal Dublin Society 7, 1845.

123 *Prince Albert, who had recently*: Henry Fitz-Patrick Berry, *A History of the Royal Dublin Society* (London and New York: Longmans, Green and Co., 1915), 279.

124 *On March 23, as Albert*: Kelly, *The Graves Are Walking*, 100.

124 *A London paper reported*: London Daily News, April 17, 1846, p. 3.

124 *He explained to a rapt audience*: Ibid.

124 *"But if this be a fact"*: Edmund Saul Dixon, *Ornamental and Domestic Poultry: Their History and Management* (London: At the Office of the Gardeners' Chronicle, 1848), 167.

124 *The birds "created such"*: Walter B. Dickson et al., *Poultry: Their Breeding, Rearing, Diseases, and General Management* (London: Henry G. Bohn, York Street, Covent Garden, 1853), 5.

124 *He was impressed by the*: James Joseph Nolan and William Oldham, *Ornamental, Aquatic, and Domestic Fowl, and Game Birds: Their Importation, Breeding, Rearing, and General Management* (Dublin: Published by the Author, at 33, Bachelor's -Walk and to Be Had of All Booksellers, 1850), 4.

125 *In the decade between 1845*: Geo P. Burnham, *The History of the Hen Fever: A Humorous Record* (Boston: J. French, 1855).

125 *"Events which are injurious"*: *The Poultry Book for the Many: Giving Full Directions for the Selection, Breeding . . . of Every Description of Poultry; with Portraits of the Principal Varieties and Plans of Poultry Houses . . . By Contributors to "the Cottage Gardener and Poultry Chronicle"* (London: Wincester 1857), 170.

126 *In 1854, Watts exchanged*: Lewis Wright, *The Book of Poultry; with Practical Schedules for Judging, Constructed from Actual Analysis of the Best Modern Decisions* (London: Cassell, 1891), 445.

126 *"If you travel by a railway"*: *The Poultry Book for the Many*, 4.

126 *"The eggs having been freely distributed"*: Dickson, *Poultry*, 10.

127 *At the time, Watts's publication noted*: Wright, *Book of Poultry*, 208.

127 *At London's popular summer*: *Poultry Chronicle*, 65.

127 *"People really seemed going"*: Wright, *Book of Poultry*, 209.

127 *The Times was disturbed by*: "The Birmingham Cattle Show," *The Times of London*, December 14, 1853.

128 *"They have fallen in"*: *Poultry Chronicle*, 66.

128 *"If the Cochin China breed"*: *The Times of London*, reprinted in *The Southern Cultivator*, (J. W. & W. S. Jones, 1853), vol. 11, 126.

7. The Harlequin's Sword

129 *"No one comes to see"*: Joanne Cooper, interview by Andrew Lawler, 2013.

129 *Now public, the museum*: "History of the Collections," Natural History Museum at Tring, accessed March 20, 2014, http://www.nhm.ac.uk/tring/history-collections /history-of-the-collections/.

130 *His 1859 book* On the Origin of Species: Charles Darwin, *The Annotated Origin: A Facsimile of the First Edition of* On the Origin of Species, annotated by James T. Costa (Cambridge, MA: Belknap Press of Harvard University Press), 2009.

131 *"God created, Linnaeus"*: David S. Kidder and Noah D. Oppenheim, *The Intellectual Devotional Biographies: Revive Your Mind, Complete Your Education, and Acquaint Yourself with the World's Greatest Personalities* (New York: Rodale, 2010), 206.

132 *Within a species, he pointed*: Georges Louis Leclerc Buffon, *Buffon's Natural History: Containing a Theory of the Earth, a General History of Man, of the Brute Creation, and of Vegetables, Minerals, &c.* (London: Printed by J. S. Barr, Bridges-Street, Covent-Garden, 1792), 353.

132 *The varieties of chickens*: Georges Louis Leclerc Buffon, *The System of Natural History*, comps. Jan Swammerdam, R. Brookes, and Oliver Goldsmith (Edinburgh, Scotland: J. Ruthven, 1800), 256.

132 *"The cock is one of the oldest companions"*: Ibid.

132 *The French naturalist Jean-Baptiste Lamarck*: Jean-Baptiste Lamarck, *Zoological Philos-*

ophy: An Exposition with Regard to the Natural History of Animals, first English trans. (New York: Hafner, 1914).

132 *In a clever stunt that drew*: Wietske Prummel et al., *Birds in Archaeology: Proceedings of the 6th Meeting of the ICAZ Bird Working Group in Groningen (23.8–27.8.2008)* (Eelde, Netherlands: Barkhuis, 2010), 279.

133 *Yet evidence mounted*: Jan van Tuyl, *A New Chronology for Old Testament Times: With Solutions to Many Hitherto Unsolved Problems through the Use of Rare Texts* (self-published via AuthorHouse, 2012), 434.

133 *The Philadelphia Quaker and physician*: Prideaux John Selby, *The Annals and Magazine of Natural History including Zoology, Botany and Geology* (London: Taylor & Francis, 1858), 211.

133 *The Charleston pastor closely*: John Bachman, *The Doctrine of the Unity of the Human Race Examined on the Principles of Science* (Charleston, SC: C. Canning, 1850), 88.

133 *An ardent opponent*: Adrian Desmond and James R. Moore, *Darwin's Sacred Cause: How a Hatred of Slavery Shaped Darwin's Views on Human Evolution* (Boston: Houghton Mifflin Harcourt, 2009), 90.

134 *He once paid for lessons*: William E. Phipps, *Darwin's Religious Odyssey* (Harrisburg, PA: Trinity Press International, 2002), 22.

134 *Listed in a contemporary directory*: Laurence S. Lockridge et al., eds., *Nineteenth-Century Lives: Essays Presented to Jerome Hamilton Buckley* (Cambridge: Cambridge University Press, 1989), 97.

135 *Wallace independently came*: Alfred Russell Wallace, *Writings on Evolution, 1843–1912*, ed. Charles H. Smith (Bristol, U.K.: Thoemmes Continuum), 2004.

135 *The Reverend Edmund Saul Dixon*: "Edmund Saul Dixon," Dickens Journals Online, accessed March 20, 2014, http://www.djo.org.uk/indexes/authors/mr-edmund-saul-dixon.html.

135 *Dixon married and moved*: Edmund Saul Dixon, *Ornamental and Domestic Poultry: Their History and Management* (London: At the Office of the Gardeners' Chronicle, 1850), viii.

136 *Dixon was puzzled*: Ibid.

136 *In 1844, an anonymous*: Robert Chambers, *Vestiges of the Natural History of Creation* (New York: W. H. Coyler), 1846.

136 *Prince Albert read it to*: Johnathan Sperber, *Europe 1850–1914: Progress, Participation and Apprehension* (Harlow, England: Pearson Longman, 2009), 46.

136 *But it also drew critics*: Desmond and Moore, *Darwin's Sacred Cause*, 218.

136 *In a gossipy letter*: "Darwin, C. R. to Hooker, J. D.," Darwin Correspondence Database, October 6, 1848, accessed March 20, 2014, http://www.darwinproject.ac.uk/letter/entry-1202.

136 *Two months later, Dixon*: Darwin's personal copy of Dixon's *Ornamental Poultry*, accessed May 14, 2014, http://www.biodiversitylibrary.org/item/106252#page/4/mode/1up.

136 *Monarchies tottered*: Karl Marx and Friedrich Engels, *The Communist Manifesto* (London: Vintage Books, 2010).

137 *On Christmas Day, as*: *Illustrated London News*, December 23, 1848.

137 *It also was*: Dixon, *Ornamental and Domestic Poultry*, x.

137 *Darwin underscored Dixon's strong*: Darwin's personal copy of Dixon's *Ornamental Poultry*.

138 *"It is very good & amusing"*: "Darwin, C. R. to Thompson, William (a)," Darwin

Correspondence Database, March 1, 1849, accessed March 20, 2014, http://www
.darwinproject.ac.uk/entry-1232.

138 *"I now really hope that your recovery"*: "Dixon, E. S. to Darwin, C. R.," Darwin Correspondence Database, April–June 1849, accessed March 20, 2014, http://www.darwin
project.ac.uk/entry-13801.

138 *Dixon grew shrill*: Desmond and Moore, *Darwin's Sacred Cause*, 219.

138 *In 1851, he declared*: Edmund Saul Dixon, *The Dovecote and the Aviary: Being
Sketches of the Natural History of Pigeons and Other Domestic Birds in a Captive State,
With Hints for Their Management* (London: Wm. S. Orr, 1851), 73.

138 *Darwin kept quiet publicly*: "Darwin, C. R. to Fox, W. D.," Darwin Correspondence
Database, July 31, 1855, accessed March 20, 2014, http://www.darwinproject.ac.uk/
entry-1733.

138 *"I was so ignorant"*: A. W. D. Larkum, *A Natural Calling: Life, Letters and Diaries of
Charles Darwin and William Darwin Fox* (Dordrecht, Netherlands: Springer, 2009), 237.

139 *Fox agreed to raise the birds*: "Darwin, C. R. to Fox, W. D.," Darwin Correspondence
Database, March 19, 1855, accessed March 20, 2014, http://www.darwinproject.ac.uk
/entry-1651.

139 *Edward Blyth, a magnificently*: "Blyth, Edward to Darwin, C. R.," Darwin Correspondence Database, August 4, 1855, accessed March 20, 2014, http://www.darwinproject
.ac.uk/entry-1735.

139 *Blyth, a brilliant but*: Robert J. Richards, *Darwin and the Emergence of Evolutionary
Theories of Mind and Behavior* (Chicago: University of Chicago Press, 1987), 107.

139 *This adaptable bird*: "Blyth, Edward to Darwin, C. R.," Darwin Correspondence Database,
September 30, 1855 or October 7, 1855, accessed March 20, 2014, http://www.darwin
project.ac.uk/entry-1761.

140 *In December 1855*: "Blyth, Edward to Darwin, C. R.," Darwin Correspondence Database,
December 18, 1855, accessed March 20, 2014, http://www.darwinproject.ac.uk/entry
-1792.

140 *While prowling a poultry*: "Darwin, C. R. to Tegetmeier, W. B.," Darwin Correspondence
Database, August 31, 1855, accessed March 20, 2014, http://www.darwinproject.ac.uk/
entry-1751.

140 *He importuned a missionary*: "Darwin, C. R. to Unspecified," Darwin Correspondence
Database, December 1855, accessed March 20, 2014, http://www.darwinproject
.ac.uk/entry-1812.

140 *In March 1856*: "Darwin, C. R. to Covington, Syms," Darwin Correspondence Database, March 9, 1856, accessed March 20, 2014, http://www.darwinproject.ac.uk/entry
-1840.

140 *"This morning I have"*: "Darwin, C. R. to Fox, W. D.," Darwin Correspondence Database,
March 15, 1856, accessed March 20, 2014, http://www.darwinproject.ac.uk/entry-1843.

140 *"By the way,"*: "Darwin, C. R. to Tegetmeier, W. B.," Darwin Correspondence Database,
October 15, 1856, accessed March 20, 2014, http://www.darwinproject.ac.uk/entry
-1975.

140 *"I expect soon"*: "Darwin, C. R. to Tegetmeier, W. B.," Darwin Correspondence Database,
November 3, 1856, accessed March 20, 2014, http://www.darwinproject.ac.uk/entry
-1981.

141 *By November, Darwin*: "Darwin, C. R. to Tegetmeier, W. B.," Darwin Correspondence
Database, January 14, 2856, accessed March 20, 2014, http://www.darwinproject
.ac.uk/entry-1820.

141 The Variation of Animals: Charles Darwin, *Charles Darwin's Works* (New York: D. Appleton, 1915).

141 *Love of novelty led*: Ibid., 240.

142 *The occasional bird*: Ibid., 238.

142 *Darwin summarizes that*: Ibid., 240.

142 *The* Gallus gallus bankiva: Ibid., 244.

142 *He ruefully acknowledges*: Ibid., 254.

142 *"Hence it may be safely"*: Ibid., 257.

143 *As recently as 2008*: Uppsala University, "Darwin Was Wrong About Wild Origin of the Chicken, New Research Shows," *ScienceDaily*, March 3, 2008, accessed March 21, 2014, http://www.sciencedaily.com/releases/2008/02/080229102059.htm.

143 *A team that included Leif*: Jonas Eriksson et al., "Identification of the Yellow Skin Gene Reveals a Hybrid Origin of the Domestic Chicken," *PLoS Genetics* 4, no. 2 (2005): E10, doi:10.1371/journal.pgen.1000010.

144 *Like his father*: "His Imperial Highness Prince Akishino (Akishinonomiya Fumihito), President, Yamashina Institute for Ornithology," Yamashina Institute for Ornithology, accessed March 21, 2014, http://www.yamashina.or.jp/hp/english/about_us/president.html.

144 *While Hirohito specialized*: Naoko Shibusawa, *America's Geisha Ally: Reimagining the Japanese Enemy* (Cambridge, MA: Harvard University Press, 2006), 104.

145 *Brisbin noted that the red*: I. Lehr Brisbin, interview by Andrew Lawler, 2012.

145 *In 2006, a team led by Yi-Ping*: Yi-Ping Liu et al., "Multiple Maternal Origins of Chickens: Out of the Asian Jungles," *Molecular Phylogenetics and Evolution*, 38, no. 1 (February 2006): 12–19.

145 *A 2012 study that used*: Alice A. Storey et al., "Investigating the Global Dispersal of Chickens in Prehistory Using Ancient Mitochondrial DNA Signatures," *PLoS ONE* 7, no. 7 (2012): E39171, doi:10.1371/journal.pone.0039171.

145 *"Earlier I thought domestication of the chicken"*: Interview with a person familiar with Prince Fumihito's views by Andrew Lawler, 2012.

145 *Olivier Hanotte, a biologist*: Olivier Hanotte, interview by Andrew Lawler, 2012.

147 *"There does not seem"*: John Lawrence et al., *Moubray's Treatise on Domestic and Ornamental Poultry: A Practical Guide to the History, Breeding, Rearing, Feeding, Fattening, and General Management of Fowls and Pigeons* (London: Arthur Hall, Virtue, and Co., 1854), 2.

147 *The Palaung people scattered*: Frederick J. Simoons, *Eat Not This Flesh: Food Avoidances from Prehistory to the Present* (Madison: University of Wisconsin Press, 1994), 145.

147 *Likewise, the Karen*: Ibid., 146.

147 *The Purum Kukis*: Ibid.

147 *At Uppsala University*: Leif Andersson, interview by Andrew Lawler, 2012.

8. The Little King

150 *A healthy rooster can produce*: Dev Raj. Kana and P. R. Yadav, *Biology of Birds* (New Delhi: DPH, Discovery Pub. House, 2005), 94.

150 *A team overseen by Martin Cohn*: Ana M. Herrera et al., "Developmental Basis of Phallus Reduction during Bird Evolution," *Current Biology* 23, no. 12 (2013): 1065–74, doi:10.1016/j.cub.2013.04.062.

150 *Cohn thinks this*: Martin Cohn, interview by Andrew Lawler, 2012.

151 *"Genitalia, dear readers,"*: Patricia Brennan, "Why I Study Duck Genitalia," *Slate*, April

2, 2013, accessed March 21, 2014, http://www.slate.com/articles/health_and_science /science/2013/04/duck_penis_controversy_nsf_is_right_to_fund_basic_research _that_conservatives.html.

151 *New England Puritans excised* cock: Toni-Lee Capossela, *Language Matters: Readings for College Writers* (Fort Worth: Harcourt Brace College Publishers, 1996), 216.

151 *Two centuries before*: Joseph Glaser, *Middle English Poetry in Modern Verse* (Indianapolis: Hackett Pub., 2007), 215.

151 *The 1785 Classical Dictionary*: Francis Grose, *A Classical Dictionary of the Vulgar Tongue, 1785* (Menston York, U.K.: Scolar P., 1968), C.

152 *The term derives*: Stewart Edelstein, *Dubious Doublets: A Delightful Compendium of Unlikely Word Pairs of Common Origin, from Aardvark/Porcelain to Zodiac/Whiskey* (Hoboken, NJ: J. Wiley, 2003), 86.

152 *"Victoria was not crowned"*: H. L. Mencken, *The American Language: An Inquiry into the Development of English in the United States* (New York: Alfred A. Knopf, 1936), 301.

152 *Well into the Victorian era*: J. Chamizo Domínguez Pedro, *Semantics and Pragmatics of False Friends* (New York: Routledge, 2008), 100.

152 *The cock most likely acquired*: Mencken, *The American Language*, 301.

152 *During separate tours*: Aharon Ben-Ze'ev, *The Subtlety of Emotions* (Cambridge, MA: MIT Press, 2000), 430.

152 *The Babylonian Talmud*: Menachem M. Brayer, *The Jewish Woman in Rabbinic Literature* (Hoboken, NJ: Ktav Pub. House, 1986), 74.

152 *The Greek god Zeus gave the handsome*: James N. Davidson, *The Greeks and Greek Love: A Bold New Exploration of the Ancient World* (New York: Random House, 2007), 223.

152 *There is no mistaking*: Lorrayne Y. Baird, "Priapus Gallinaceus: The Role of the Cock in Fertility and Eroticism in Classical Antiquity and the Middle Ages," *Studies in Iconography* 7–8 (1981–82): 81–111.

153 *"Throughout classical antiquity"*: Ibid.

153 *In classical art, roosters*: Exhibit in Altes Museum, Berlin, personal visit by Andew Lawler, 2013.

153 *Hidden from public view*: Baird, "Priapus Gallinaceus."

153 *"In Nature man generates"*: Scott Atran, *Cognitive Foundations of Natural History: Towards an Anthropology of Science* (Cambridge: Cambridge University Press, 1990), 87.

154 *By the time of Christ*: Richard Payne Knight, *The Symbolical Language of Ancient Art and Mythologie: An Inquiry* (New York: J. W. Bouton, 1892), 150.

154 *"They declare the cock*: Ibid., 70.

154 *"Praised be to the Lord"*: Isidore Singer, *The Jewish Encyclopedia: A Descriptive Record of the History, Religion, Literature, and Customs of the Jewish People from the Earliest Times to the Present Day* (New York: Funk and Wagnalls, 1904), 3:11.

154 *"Lo, the raving lions"*: Lucretius, *On the Nature of Things*, trans. William Ellery Leonard (Sioux Falls, SD: NuVision Publications, 2007), 124.

154 *Given this wealth of tradition*: *Dictionary of Christianity*, comp., J. C. Cooper (Chicago: Fitzroy Dearborn, 1996), s.v. "Animals."

154 *These are ostensibly reminders*: Mark 14:30 (*New American Bible*).

154 *But the bird's heralding of light*: Lorrayne Y. Baird-Lange, "Christus Gallinaceus: A Chaucerian Enigma; or the Cock as Symbol of Christ in the Middle Ages," *Studies in Iconography* 9 (1983): 19–30.

154 *In a popular fourth-century AD*: James J. Wilhelm, *The Cruelest Month: Spring, Nature, and Love in Classical and Medieval Lyrics* (New Haven: Yale University Press, 1965), 63.

154 *"God, the Creator Himself"*: Baird-Lange, "Christus Gallinaceus," 21.

155 *Peter was crucified on Vatican Hill*: James George Roche Forlong, *Faiths of Man: A Cyclopedia of Religions*, vol. 2 (London: B. Quaritch, 1906), s.v. "Janus."

155 *"I will give you the keys*: Matthew 16:19 (*New American Bible*).

155 *Today's Saint Peter's*: Encyclopaedia Britannica, vols. 11–12, s.v. "Great Mother of the Gods."

155 *"The erotic association"*: Randy P. Conner et al., *Cassell's Encyclopedia of Queer Myth, Symbol, and Spirit: Gay, Lesbian, Bisexual, and Transgender Lore* (London: Cassell, 1997), s.v. "Attis."

155 *The poet Juvenal remarked*: David M. Friedman, *A Mind of Its Own* (New York: Simon & Schuster, 2008), 31.

155 *Near both the original basilica and the Cybele temple*: J. G. Heck and Spencer Fullerton Baird, comps., *Iconographic Encyclopaedia of Science, Literature, and Art* (New York: R. Garrigue, 1857), s.v. "Rome."

155 *By the late sixth century*: John G. R. Forlong, comp., *Encyclopedia of Religions* (New York: Cosimo Classics, 2008), s.v. "Peter."

155 *"The Cock is like the Souls"*: Louisa Twining, *Symbols and Emblems of Early and Medieval Christian Art* (London: John Murray, 1885), 188.

155 *By the ninth century*: Norwood Young, ed., *Handbook for Rome and the Campagna* (London: Edward Stanford, 1908), s.v. "S. Pietro."

155 *Clergy were called*: Hargrave Jennings, *Phallicism: Celestial and Terrestrial, Heathen and Christian* (London: George Redway, 1884).

156 *By the tenth century*: Forlong, *Faiths of Man*, s.v. "Peter."

156 *In 1102, at the end of the First*: Menashe Har-El, *Golden Jerusalem* (Jerusalem: Gefen Pub. House, 2004), 311.

156 *By the time of the First Crusade*: Baird-Lange, "Christus Gallinaceus," 26.

156 *Magical amulets displaying fierce creatures*: Joseph Campbell, *The Masks of God* (New York: Viking Press, 1959), 275.

156 *In the fourteenth century*: Mark Allen, *The Complete Poetry and Prose of Geoffrey Chaucer* (Boston: Wadsworth Cengage Learning, 2011), 239.

156 *Alchemists, those protochemists*: Elio Corti, trans., "The Chicken of Ulisse Aldrovandi," accessed March 21, 2014, http://archive.org/stream/TheChickenOfUlisse Aldrovandi/Aldrogallus_djvu.txt.

156 *A last echo of the cock's Christian*: William Shakespeare, *Shakespeare's Tragedy of Hamlet*, ed. John Hunter (London: Longmans, Green and Co., 1874), 11.

157 *A decade later in the town*: Marino Niola, "Archeologia della devozione" in L. M. Lombardi Satriani, ed., *Santità e tradizione: itinerari antropologico-religiosi* in *Campania*, 2000.

157 *"Within the Cock, vile"*: Stephen Orgel, *The Authentic Shakespeare, and Other Problems of the Early Modern Stage* (New York: Routledge, 2002), 217.

157 *"What is your crest?"*: William Shakespeare, *The Complete Works of William Shakespeare* (New York: G.F. Cooledge & Brother, 1844), 262.

157 *Some bard scholars*: Joel Friedman, "The Use of the Word 'Comb' in Shakespeare's *The Taming of the Shrew* and *Cymbeline*," *Joel Friedman Shakespeare Blog*, February 9, 2010, accessed March 21, 2014, http://joelfriedmanshakespeare.blogspot .com/2010/02/use-of-word-comb-in-shakespeares-taming.html.

158 *Thirty years later, citing*: "The Ceremony of Presenting a Cock to the Pope," *American Ecclesiastical Review* 29 (1903): 301.

158 *The Puritan preacher Cotton Mather*: *The Congregationalist and Christian World* 100 (1915): 156.

158 *The rooster remains*: France in the United States/Embassy of France in Washington, "The Gallic Rooster," December 20, 2013, accessed March 21, 2014 http://www.amba france-us.org/spip.php?article604.

158 *It also has a longer history*: Steven Seidman, "The Rooster as the Symbol of the U.S. Democratic Party," *Posters and Election Propaganga* (blog), Ithaca College, June 12, 2010, accessed March 21, 2014, http://www.ithaca.edu/rhp/programs/cmd/blogs/posters _and_election_propaganda/the_rooster_as_the_symbol_of_the_u.s._democratic_p/# .Uywtbl76'Ivo.

158 *In 2007, a scientific team*: J. M. Asara et al., "Protein Sequences from Mastodon and Tyrannosaurus Rex Revealed by Mass Spectrometry," *Science* 316, no. 5822 (2007): 280–85, doi:10.1126/science.1137614.

158 "*Study: Tyrannosaurus*": Jeanna Bryner, "Study: *Tyrannosaurus Rex* Basically a Big Chicken," *Fox News*, April 25, 2008, accessed March 21, 2014, http://www.foxnews .com/story/2008/04/25/study-tyrannosaurus-rex-basically-big-chicken/.

159 *The discovery began*: Evan Ratliff, "Origin of Species: How a *T. Rex* Femur Sparked a Scientific Smackdown," *Wired*, June 22, 2009, accessed May 14, 2014, http://archive .wired.com/medtech/genetics/magazine/17-07/ff_originofspecies?currentPage=all.

159 *A Harvard University chemist*: John Asara, interview by Andrew Lawler, 2013.

160 *Their 2007 paper in*: M. H. Schweitzer et al., "Biomolecular Characterization and Protein Sequences of the Campanian Hadrosaur B. Canadensis," *Science* 324, no. 5927 (2009): 626–31, doi:10.1126/science.1165069.

160 *Horner, the Montana paleontologist*: Jack Horner, "Jack Horner: Building a Dinosaur from a Chicken," TED video, 16:36, talk presented at an official TED conference, March 2011, accessed March 21, 2014, http://www.ted.com/talks/jack_horner_building_a _dinosaur_from_a_chicken.

161 *In 2004, a biologist*: Matthew P. Harris et al., "The Development of Archosaurian First-Generation Teeth in a Chicken Mutant," *Current Biology* 16, no. 4 (2006): 371–77, doi:10.1016/j.cub.2005.12.047.

161 *Arkhat Abzhanov, an evolutionary*: Arkhat Abzhanov, interview by Andrew Lawler, 2013.

162 *In 2007, paleontologists*: A. H. Turner et al., "Feather Quill Knobs in the Dinosaur Velociraptor," *Science* 317, no. 5845 (2007): 1721, doi:10.1126/science.1145076.

162 *Four years later, chunks*: R. C. McKeller et al., "A Diverse Assemblage of Late Cretaceous Dinosaur and Bird Feathers from Canadian Amber," *Science* 333, no. 6049 (2011): 1619–622, doi:10.1126/science.1203344.

162 *Triceratops, that hulking*: J. Clarke, "Feathers Before Flight," *Science* 340, no. 6133 (2013): 690-92. doi:10.1126/science.1235463.

163 *Feathers, in fact, may*: Ibid.

163 *Avians are a particular sort*: Gareth Dyke and Gary W. Kaiser, *Living Dinosaurs: The Evolutionary History of Modern Birds* (Chichester, West Sussex: Wiley-Blackwell, 2011), 9.

163 *Some researchers even*: "Oviraptor," Mediahex, accessed March 21, 2014, http://www .mediahex.com/Oviraptor.

163 *Called a basilisk*: Rosa Giorgi et al., *Angels and Demons in Art* (Los Angeles: J. Paul Getty Museum, 2005), 99.

163 *Pliny the Elder*: Ibid.

164 *The twelfth-century German mystic*: Petzoldt Leander, *Kleines Lexikon der Dämonen und Elementargeister* (Munich: C. H. Beck, 2003), 30.

164 *Panic swept thirteenth-century Vienna*: George M. Eberhart, *Mysterious Creatures: A Guide to Cryptozoology* (Santa Barbara, CA: ABC-CLIO, 2002), 33.

164 *A hysterical mob of sixteenth-century Dutch*: Mike Dash, "The Strange Tale of the Warsaw Basilisk," *A Fortean in the Archives* (blog), February 28, 2010, accessed March 21, 2014, http://aforteantinthearchives.wordpress.com/2010/02/28/the-strange-tale-of-the-warsaw-basilisk/.

164 *On an August afternoon in 1474*: Thomas Hofmeier, *Basils Ungeheuer: Eine kleine Basiliskenkunde* (Berlin and Basel: Leonhard-Thurneysser-Verlag, 2009).

164 *As late as 1651*: Jan Bondeson, *The Feejee Mermaid and Other Essays in Natural and Unnatural History* (Ithaca: Cornell University Press, 1999), 176.

164 *"Know, Sir, that there"*: Voltaire, *Zadig, Or, The Book of Fate: An Oriental History; Translated from the French, Etc.* (London: T. Kelly, 1816), 48.

164 *The basilisk has resurfaced*: Allan Zola Kronzek and Elizabeth Kronzek, *The Sorcerer's Companion: A Guide to the Magical World of Harry Potter* (New York: Broadway Books, 2001), 22.

165 *Clinton wanted to be a*: Mike Clinton, interview by Andrew Lawler, 2012.

165 *The Greek philosopher*: Georgios Anagnostopoulos, *A Companion to Aristotle* (Chichester, U.K.: Wiley-Blackwell, 2009), 113.

167 *Clinton's surprising conclusion*: D. Zhao et al., "Somatic Sex Identity Is Cell Autonomous in the Chicken," *Nature* 464, no. 7286 (2010): 237–42, doi:10.1038/nature08852.

168 *"Successful chicken-sexing"*: Lyall Watson, *Beyond Supernature: A New Natural History of the Supernatural* (Toronto: Bantam Books, 1988), 65.

168 *In the United States alone, more*: Associated Press, "Chicks Being Ground Up Alive Video," *The Huffington Post*, September 1, 2009, accessed March 21, 2014, http://www.huffingtonpost.com/2009/09/01/chicks-being-ground-up-al_n_273652.html.

168 *For such sex detection*: Clinton, interview.

9. Feeding Babalu

170 *There is a day when*: Leon Rubin and I. Nyoman Sedana, *Performance in Bali* (London: Routledge, 2007), 4.

171 *Ida Pedanda Made Manis*: Ida Pedanda Made Manis, interview by Andrew Lawler, 2013.

171 Pedanda *means "bearer"*: Tom Hunter Aryati, translations via email message to author, May 13, 2014.

172 *Though they enjoy a reputation*: Hugh Mabbett, *The Balinese* (Wellington, N.Z.: January Books, 1985), 97.

172 *A Balinese king told one*: J. Stephen Lansing, *Perfect Order: Recognizing Complexity in Bali* (Princeton, N.J.: Princeton University Press, 2012), 56.

172 *The focus today is on animal*: Leo Howe, *The Changing World of Bali: Religion, Society and Tourism* (London: Routledge, 2005).

172 *Terrorist bombings at a packed*: "The 12 October 2002 Bali Bombing Plot," *BBC News*, October 11, 2012, accessed March 21, 2014, http://www.bbc.com/news/world-asia-19881138.

172 *To redress the imbalance*: Richard C. Paddock, "Prayerful Balinese Gather to Purge

Bombing Site of Evil," *Los Angeles Times*, November 16, 2002, accessed March 21, 2014, http://articles.latimes.com/2002/nov/16/world/fg-bali16.

172 *That ritual paled in comparison*: Cameron Forbes, *Under the Volcano: The Story of Bali* (Melbourne, Vic.: Black, 2007), 76.

172 *"Night and day"*: Tom Hunter Aryati, translations via email message to author, March 10, 2012.

173 *Pura Penataran Agung Taman Bali*: Ibid.

175 *"We appear to substitute"*: Yancey Orr, interview by Andrew Lawler, 2013.

175 *Technically, only three matches*: "Cockfighting: Cockfighting in Indonesia," indahnesia .com, last modified September 15, 2009, accessed March 21, 2014, http://indahnesia .com/indonesia/INDCOC/cockfighting.php.

175 *"To anyone who"*: Clifford Geertz, *The Interpretation of Cultures: Selected Essays* (New York: Basic Books, 1973), 417.

176 *"Children of men who"*: John A. Hardon, *American Judaism* (Chicago: Loyola University Press, 1971), 179.

177 *The Roman Jewish writer*: Ronald L. Eisenberg, *The JPS Guide to Jewish Traditions* (Philadelphia: Jewish Publication Society, 2004), 713.

177 *"How often I have longed"*: Matthew. 23:37 (*The New Jerusalem Bible: Standard Edition*).

177 *The Mishnah, an older part*: Eisenberg, *The JPS Guide*, 712.

177 *A later portion of the Talmud*: Ronald L. Eisenberg, *What the Rabbis Said: 250 Topics from the Talmud* (Santa Barbara, CA: Praeger, 2010), 266.

177 *The Hebrew word for rooster, gever*: Joshua Trachtenberg, *Jewish Magic and Superstition: A Study in Folk Religion* (Philadelphia: University of Pennsylvania Press, 2004), 164.

177 *The practice of kapparot*: Adele Berlin, *The Oxford Dictionary of the Jewish Religion* (New York: Oxford University Press, 2011), s.v. "Kapparot"; Nancy E. Berg, *Exile from Exile: Israeli Writers from Iraq* (Albany: State University of New York Press, 1996), 60.

178 *Thirteenth-century rabbis*: Israel Drazin, *Maimonides: The Exceptional Mind* (Jerusalem: Gefen Pub., 2008), 203.

179 *In 2005, a chicken vendor*: Tanay Warerkar and Oren Yaniv, "No One Here but Us (Dead) Chickens! Thousands of Birds Die from Heat, Not Jewish Sin Ritual," *New York Daily News*, September 12, 2003, accessed March 21, 2014, http://www.nydaily news.com/new york/brooklyn/birds-die-animal-ritual-slaughter article 1.1151098.

179 *"The industrialization"*: Associated Press, "Activists Cry Foul over Ultra-Orthodox Chicken Ritual," *NewsOK*, September 8, 2010, accessed March 21, 2014, http://new sok.com/activists-cry-foul-over-ultra-orthodox-chicken-ritual/article/feed/189277.

180 *"Kapparot is not consistent"*: Goren quoted by Nazila Mahgerefteh, "A Wing and a Prayer," September 28, 2006, accessed May 15, 2014, www.endchickensaskaporos.com.

180 *The day before my visit*: Rachel Avraham, "Minister Peretz Opposes the Kaparot Tradition: 'Prevent Animals from Suffering,'" JerusalemOnline, September 11, 2013, accessed March 21, 2014, http://www.jerusalemonline.com/news/politics-and -military/politics/minister-peretz-opposes-the-kaparot-tradition-1616.

180 *Hundreds of fowl slated for use*: Warerkar and Yaniv, "No One Here but Us (Dead) Chickens!"

180 *James Frazer noted*: James George Frazer and Robert Fraser, *The Golden Bough: A Study in Magic and Religion* (Oxford: Oxford University Press, 2009), 14.

180 *Frazer notes that*: Ibid., 545.

181 *"You may kill animals"*: *Church of Lukumi Babalu Aye, Inc. v. City of Hialeah*, 508 U.S. 520 (1993).

181 *Drawing on ancient African*: Abiola Irele and Biodun Jeyifo, *The Oxford Encyclopedia of African Thought* (New York: Oxford University Press, 2010), s.v. "Santeria."

181 *"What can we do"*: Ronald J. Krotoszynski, *The First Amendment: Cases and Theory* (New York: Aspen Publishers, 2008), 881.

182 Lukumi *is another term for Santeria*: George Brandon, *Santeria from Africa to the New World: The Dead Sell Memories* (Bloomington: Indiana University Press, 1993), 145.

182 *"Babalu," Ricky Ricardo's*: Migene González-Wippler, *Santeria: The Religion, Faith, Rites, Magic* (St. Paul, MN: Llewellyn Worldwide, 1994), 68.

182 *White hens and roosters*: Caracas vendor who declined to give name, interview by Andrew Lawler, 2013.

182 *About AD 1200, a settlement*: Toyin Falola and Christian Jennings, *Sources and Methods in African History: Spoken, Written, Unearthed* (Rochester, NY: University of Rochester Press, 2003), 59.

182 *According to one Yoruba tradition*: Abraham Ajibade Adeleke, preface to *Intermediate Yoruba: Language, Culture, Literature, and Religious Beliefs*, part 2 (Bloomington, IN: Trafford Pub, 2011).

183 *"Such a name for"*: Daniel McCall, "The Marvelous Chicken and Its Companion in Yoruba Art," *Paideuma* Bd. 24 (1978).

183 *The first mention of the bird*: Timothy Insoll, *The Archaeology of Islam in Sub-Saharan Africa* (Cambridge, U.K.: Cambridge University Press, 2003), 239; R. Blench and Kevin C. MacDonald, *The Origins and Development of African Livestock: Archaeology, Genetics, Linguistics, and Ethnography* (London: UCL Press, 2000).

183 *In a 2011 Ethiopian dig*: Catherine D'Andrea, email message to author, 2013.

184 *Stephen Dueppen, a professor*: Stephen Dueppen, interview by Andrew Lawler, 2012; see Stephen A. Dueppen, *Egalitarian Revolution in the Savanna: The Origins of a West African Political System* (London: Equinox Pub., 2012).

186 *Among neighboring Mali*: Jean Chevalier et al., *A Dictionary of Symbols* (London: Penguin Books, 1996), s.v. "Rooster."

186 *In his 2010 paper in American Antiquity*: S. A. Dueppen, "Early evidence for chicken at Iron Age Kirikongo (c. AD 100–1450), Burkina Faso," *Antiquity* 85, no. 327: 142–57.

186 *In the Congo basin, a female*: Victor Turner, "Poison Ordeal: Revelation and Divination in Ndembu Ritual," Object Retrieval, accessed March 21, 2014, http://www.objectretrieval.com/node/273.

186 *Yoruba proverbs are liberally*: Oyekan Owomoyela, *Yoruba Proverbs* (Lincoln: University of Nebraska Press, 2005), 467.

187 *Before any major undertaking*: Federico Santangelo, *Divination, Prediction and the End of the Roman Republic* (Cambridge: Cambridge University Press, 2013), 27.

188 *"Scarcely any matter out"*: Marcus Tullius and Clinton Walker Keyes, eds., Cicero, *De Re Publica, De Legibus* (Cambridge, MA: Harvard University Press, 1951), 255.

188 *A specialist called a* pullarius: Santangelo, *Divination, Prediction, and the End of the Roman Republic*, 27.

188 *The sacred chickens kept*: Cicero, *De Re Publica, De Legibus*, 393.

188 *A senior Roman general*: Ibid., 451.

189 *"I think that, although"*: Ibid., 226.

189 *"In a world of instant"*: Ócha'ni Lele, *Teachings of the Santeria Gods: The Spirit of the Odu* (Rochester, VT: Bear & Co., 2010), 7.

190 *Testifying to the U.S. district*: Church of the Lukumi Babalu Aye v. City of Hialeah.

190 *Sanitary workers at the Miami-Dade*: "Miami Courthouse Littered with Sacrifices to Gods," *New York Daily News*, April 10, 1995, accessed March 21, 2014, http://news.google.com/newspapers?nid=1241&dat=19950410&id=AZ1YAAAAIBAJ&sjid=G4YDAAAAIBAJ&pg=6744,5351003.

190 *In June 1993, all nine*: Church of the Lukumi Babalu Aye v. City of Hialeah.

10. Sweater Girls of the Barnyard

192 *Food grew so short in 1610*: Rachel Herrmann, "The 'Tragicall Historie': Cannibalism and Abundance in Colonial Jamestown," *William and Mary Quarterly*, 3rd ser., 68, no. 1 (January 2011).

192 *In New England, where chickens arrived*: Keith W. F. Stavely and Kathleen Fitzgerald, *Northern Hospitality: Cooking by the Book in New England* (Amherst: University of Massachusetts Press, 2011), 185.

192 *Grateful for the exotic*: A. R. Hope Moncrieff, *The Heroes of Young America* (London: Stanford, 1877), 221.

193 *On Winslow's farm, wild*: "Archaelogy of the Edward Winslow Site," http://www.plymoutharch.com/.

193 *"Seventeenth- and eighteenth-century"*: Andrew F. Smith, ed., *The Oxford Encyclopedia of Food and Drink in America* (Oxford: Oxford University Press, 2004), s.v. "Chicken Cookery."

193 *In 1692, after several*: Patricia A. Gibbs, "Slave Garden Plots and Poultry Yards," Colonial Williamsburg, accessed March 21, 2014, http://research.history.org/historical_research/research_themes/themeenslave/slavegardens.cfm.

193 *The chicken "is the only"*: Eugene Kusielewicz, Ludwik Krzyżanowski, "Julian Ursyn Niemcewicz's American Diary," *The Polish Review* 3 (Summer 1958): 102.

193 *When Washington ordered*: "From George Washington to Anthony Whitting, 26 May 1793," Washington Papers, Founders Online, accessed March 21, 2014, http://founders.archives.gov/documents/Washington/05-12-02-0503.

193 *Virginia planter Landon Carter*: Psyche A. Williams-Forson, *Building Houses Out of Chicken Legs: Black Women, Food, and Power* (Chapel Hill: University of North Carolina Press, 2006), 16.

194 *As early as 1665, Maryland*: James D. Rice, *Nature & History in the Potomac Country: From Hunter-Gatherers to the Age of Jefferson* (Baltimore: Johns Hopkins University Press, 2009), 136.

194 *A century later*: James Mercer to Battaile Muse, April 3, 1779, Battaile Muse Papers, ed. John C. Fitzpatrick, William R Perkins Library, Duke University.

194 *In one letter to his overseer*: Philip D. Morgan, *Slave Counterpoint: Black Culture in the Eighteenth-Century Chesapeake and Lowcountry* (Chapel Hill: published for the Omohundro Institute of Early American History and Culture, Williamsburg, Virginia, by the University of North Carolina Press, 1998), 364.

194 *In what likely was a typical*: "Economy," Landscape of Slavery: Mulberry Row at Monticello, accessed March 21, 2014, http://www.monticello.org/mulberry-row/topics/economy.

194 *Chickens and eggs were*: Gerald W. Gawalt, "Jefferson's Slaves: Crop Accounts at Monticello, 1805–1808," *Journal of the AfroAmerican Historical and Genealogical Society*, Spring/Fall 1994, 19–20.

194 *In 1728, a white owner named*: Morgan, *Slave Counterpoint*, 361.

194 *"Adjoining their little habitations"*: John P. Hunter, *Link to the Past, Bridge to the Future: Colonial Williamsburg's Animals* (Williamsburg, VA: Colonial Williamsburg Foundation, 2005), 50.

194 *One traveler passing*: Morgan, *Slave Counterpart*, 370.

194 *In Charleston, black women*: Williams-Forson, *Building Houses Out of Chicken Legs*, 24.

195 *"The slaves sell eggs"*: Fredrika Bremer and Mary Botham Howitt, *The Homes of the New World: Impressions of America* (New York: Harper & Bros., 1853), 297.

195 *Chicken-rich slaves on one*: Billy G. Smith, *Down and Out in Early America* (University Park: Pennsylvania State University Press, 2004), 113.

195 *In one South Carolina rhyme*: Julia Floyd Smith, *Slavery and Rice Culture in Low Country Georgia, 1750–1860* (Knoxville: University of Tennessee Press, 1985), 176.

195 *In a deposition, Prosser*: Harry Kollatz, *True Richmond Stories: Historic Tales from Virginia's Capital* (Charleston, SC: History Press, 2007), 43.

196 *Her fame as a chef spread*: Mary Randolph, *The Virginia House-wife* (Washington: printed by Davis and Force, 1824), 75.

196 *A century later, another*: Josh Ozersky, *Colonel Sanders and the American Dream* (Austin: University of Texas Press, 2012).

196 *More than ten thousand people*: John Henry Robinson, *The First Poultry Show in America, Held at the Public Gardens, Boston, Mass., Nov. 15–16, 1849: An Account of the Show Comp. from Original Sources* (Boston, MA: Farm-Poultry Pub., 1913), 8.

196 *"Everybody was there"*: Geo P. Burnham, *The History of the Hen Fever: A Humorous Record* (Boston: J. French and Co., 1855), 24.

197 *These were from the same stock*: Ibid., 16.

197 *In gratitude, the grateful monarch*: Ibid., 129.

197 *The consummate American showman*: Ibid., 194.

197 *"There will be a marvelous cackling*: "The National Poultry Show," *New York Times*, February 13, 1854, 8.

197 *He repeated the wildly successful*: Francis H. Brown, "Barnum's National Poultry Show Polka," 1850.

198 *Editors warned the public*: "The Hen Fever," *Genesee Farmer*, January 1851, 16.

198 *An upstate New York newspaper*: "A Valuable Hen," *Southern Cultivator* 11 (reprint from *Rochester Daily Advertiser*), 1853.

198 *A plantation owner in Rome, Georgia*: "Hard Fare for the Poor Negroes," *Southern Cultivator* 11 (reprint from *Northern Farmer*), 1853.

198 *In his 1853 short story*: William B. Dillingham, ed., *Melville's Short Fiction, 1853–1856* (Athens: University of Georgia Press, 1977).

198 *"A cock, more like a golden"*: Ibid., 60.

199 *The Plymouth Rock, one*: Andrew F. Smith, *The Saintly Scoundrel: The Life and Times of Dr. John Cook Bennett* (Urbana: University of Illinois Press, 1997), 168.

199 *In 1875, a Maine farmer*: B. F. Kaupp, *Poultry Culture Sanitation and Hygiene* (Philadelphia: Saunders, 1920), 37.

199 *While holding the bird sacred*: "Marcus Terentius Varro on Agriculture," accessed March 21, 2014, http://penelope.uchicago.edu/Thayer/E/Roman/Texts/Varro/de_Re _Rustica/3%2A.html.

200 *Varro recommended housing*: Ibid.

200 *"We disregard the chief purpose"*: Robert Joe Cutter, *The Brush and the Spur: Chinese Culture and the Cockfight* (Hong Kong: Chinese University Press, 1989), 141.

200 *A second-century AD stone relief*: John R. Clarke, *Art in the Lives of Ordinary Romans: Visual Representation and Non-elite Viewers in Italy, 100 B.C.–A.D. 315* (Berkeley: University of California Press, 2003), 124.

200 *The Roman Empire's only surviving*: Apicius, *Apicius: A Critical Edition with an Introduction and an English Translation of the Latin Recipe Text Apicius*, eds. C. W. Grocock and Sally Grainger (Totnes, U.K.: Prospect, 2006), 231.

201 *The temperature around*: "Incubation and Embryology Questions and Answers," University of Illinois Extension, Incubation and Embryology, accessed March 22, 2014, http://urbanext.illinois.edu/eggs/res32-qa.html.

201 *Jefferson complained in 1812*: H. A. Washington, ed., "The Writings of Thomas Jefferson," accessed March 22, 2014, http://www.yamaguchy.com/library/jefferson/1812.html.

201 *But ancient Egyptians and Chinese*: "Hatching Eggs with Incubators," from *Lessons with Questions, 1–20* (Topeka, KS: National Poultry Institute, 1914), 185.

201 *By the medieval era, Europeans*: Paulina B. Lewicka, *Food and Foodways of Medieval Cairenes: Aspects of Life in an Islamic Metropolis of the Eastern Mediterranean* (Leiden, Netherlands: Brill, 2011), 202.

201 *A Medici managed*: The editors and contributors to *The Journal of Horticulture, The Garden Manual for the Cultivation and Operations Required for the Kitchen Garden, Flower Garden, Fruit Garden, Florists' Flowers* (London: Journal of Horticulture & Home Farmer Office, 1893), 253.

202 *The French polymath*: René-Antoine Ferchault De Réaumur, *The Art of Hatching and Bringing up Domestick Fowls of All Kinds at Any Time of the Year: Either by Means of the Heat of Hot-beds, or That of Common Fire*, ed. Charles Davis (London: printed for C. Davis, 1750), 6.

202 *The resulting chicks delighted*: Bridget Travers and Fran Locher Freiman, *Medical Discoveries: Medical Breakthroughs and the People Who Developed Them* (Detroit: UXL, 1997), 247.

202 *In 1880, there were 100 million*: *Report of the Chief of the Bureau of Animal Industry*, vol. 19 (Washington, D.C.: United States Department of Agriculture, U.S. Government Printing Office, 1903).

202 *The Talmud calls for Orthodox*: Norman Solomon, *The Talmud: A Selection* (London: Penguin, 2009); Isaiah 58:13.

202 *At first, most of the city's supply*: Jay Shockley, "Gansevoort Market Historic District Designation Report," part 1 (New York City Landmarks Preservation Commission, September 9, 2003).

202 *"Two hundred thousand dozen eggs"*: "Eggs from Foreign Lands," *New York Times*, June 14, 1883, 8.

203 *One horrified rabbi wrote*: Sue Fishkoff, *Kosher Nation* (New York: Schocken Books, 2010), 58.

203 *In 1900, there were fifteen hundred*: Kenneth T. Jackson, *The Encyclopedia of New York City* (New Haven: Yale University Press, 1995), s.v. "Kosher Foods."

203 *The emigrants called the*: Williams-Forson, *Building Houses Out of Chicken Legs*, 116.

203 *Although 90 percent of North Carolina*: William S. Powell, ed., *The Encyclopedia of North Carolina* (Chapel Hill: University of North Carolina Press, 2006), s.v. "Poultry."

204 *One North Carolina extension agent*: Lu Ann Jones, *Mama Learned Us to Work: Farm Women in the New South* (Chapel Hill: University of North Carolina Press, 2002), 85.

204 *In 1909, a teenager named Mollie Tugman*: Ibid., 87.

204 *Tugman was inspired by articles*: Ibid., 99.

204 *Another Carolinian named H. P. McPherson*: Ibid., 87.

204 *Poultry was the state's fastest*: Powell, *The Encyclopedia of North Carolina*.

205 *"Save hens, raise hens, eat eggs"*: Government advertisement, *San Francisco Chronicle*, April 7, 1918.

205 *Another poster featuring a hen stated*: Government advertisement, *American Poultry Advocate* 26, 1917, 182.

205 *That March, the post office agreed*: George J. Mountney, *Poultry Products Technology*, 3rd ed. (London: Taylor & Francis, 1995), 22.

205 *One entrepreneur in California*: Joseph Tumback, *How I Made $10,000 in One Year with 4200 Hens* (n.p.: Joseph H. Tumback, 1919).

205 *The wife of a textile manufacturer*: Jones, *Mama Learned Us to Work*, 93.

205 *In Delaware, Celia Steele*: Frank Gordy, "National Register of Historic Places Inventory—Nomination Form: First Broiler House," National Park Service, 1972.

206 *In 1925, farmers on Delmarva*: Gordon Sawyer, *The Agribusiness Poultry Industry; A History of Its Development* (New York: Exposition Press, 1971), 37.

206 *A U.S. Department of*: Ibid., 46.

206 *In 1928, the Republican National*: *Webster's Guide to American History: A Chronological, Geographical, and Biographical Survey and Compendium* (Springfield, MA: Merriam, 1971), s.v. "Chronology 1928; Republican National Committee Advertisement."

206 *What had begun as*: Jones, *Mama Learned Us to Work*, 99.

206 *By then, ten thousand railroad cars*: Ibid.

207 *As a band played and the crowd*: *American Magazine* 152 (1951), 104.

207 *With* Chicken Little: Martin J. Manning and Herbert Romerstein, *Historical Dictionary of American Propaganda* (Westport, CT: Greenwood Press, 2004), s.v. "Disney Image."

207 *The agency promptly seized*: Solomon I. Omo-Osagie, *Commercial Poultry Production on Maryland's Lower Eastern Shore: The Role of African Americans, 1930s to 1990s* (Lanham, MD: University Press of America, 2012), 49.

207 *Soon chicken was standard*: Williams-Forson, *Building Houses Out of Chicken Legs*, 66.

208 *chickens were said to be*: M. B. D. Norton, *A People and a Nation: A History of the United States: Volume II: Since 1865* (Boston: Houghton Mifflin, 1986), 747.

208 *But the internment of Japanese*: Sawyer, *The Agribusiness Poultry Industry*, 77.

208 *By the time the war ended*: *Big Chicken: Pollution and Industrial Poultry Production in America* (Washington, D.C.: Pew Environment Group, 2011), 9.

208 *In a 1945 meeting in Canada*: Margaret Elsinor Derry, *Art and Science in Breeding: Creating Better Chickens* (Toronto: University of Toronto Press, 2012), 165; *American Poultry Journal: Broiler Producer Edition*, vols. 89–93 (1958): 90.

208 *Smarting from a federal conviction*: Marc Levinson, *The Great A&P and the Struggle for Small Business in America* (New York: Hill and Wang, 2011), 234.

209 *On the Chicken of Tomorrow Committee*: Karl C. Seeger, *The Results of the Chicken-of-Tomorrow 1948 National Contest* (University of Delaware, Agricultural Experiment Station, 1948).

209 *The goal was to draw on the expertise of small*: Levinson, *The Great A&P*, 241.

209 *Given the old emphasis on egg*: *The Chicken of Tomorrow*, directed by Jack Arnold, 1948, Audio Productions Inc., accessed May 14, 2014, https://archive.org/details/Chickeno1948.

209 *The committee also cosponsored*: Susan Merrill Squier, *Poultry Science, Chicken Culture: A Partial Alphabet* (New Brunswick, NJ: Rutgers University Press, 2011), 51.

209 *Contests in forty-two of the forty-eight states*: American Poultry 25 (1949).

209 *Then they were weighed*: Derry, *Art and Science in Breeding*, 165.

210 *A 1951 issue of the* Arkansas Agriculturalist: Ibid., 50.

210 *Until the early 1950s*: U.S. Dept. of Commerce, *United States Census of Agriculture, 1954* (Washington, D.C.: U.S. Dept. of Commerce, Bureau of the Census, 1956), 9.

210 *The only exception*: Alissa Hamilton, *Squeezed: What You Don't Know about Orange Juice* (New Haven: Yale University Press, 2009), 25.

210 *Advances in nutrition and*: "Production Systems," U.S. Environmental Protection Agency, accessed March 24, 2014, http://www.epa.gov/oecaagct/ag101/poultrysystems .html.

211 *John Tyson, for example*: Encyclopedia of Arkansas History and Culture, "Tyson Foods, Inc.," accessed March 22, 2014, http://www.encyclopediaofarkansas.net/encyclopedia /entry-detail.aspx?entryID=2101.

211 *By providing feed and water to the birds*: "Donald John Tyson—Overview, Personal Life, Career Details, Chronology: Donald John Tyson, Social and Economic Impact," accessed March 22, 2014, http://encyclopedia.jrank.org/articles/pages/6375 /Tyson-Donald-John.html.

211 *Only the largest operations*: Marvin Schwartz, *Tyson: From Farm to Market* (Fayetteville: University of Arkansas Press, 1991), 10.

211 *"Just keep it simple*: Christopher Leonard, *The Meat Racket: The Secret Takeover of America's Food Business* (New York: Simon & Schuster, 2014), epigraph.

211 *Tyson's son Don studied*: Schwartz, *Tyson*, 9.

211 *"We're not committed"*: Brenton Edward Riffel, *The Feathered Kingdom: Tyson Foods and the Transformation of American Land, Labor, and Law, 1930–2005* (ProQuest, 2008), 146.

212 *So the new generation*: Derry, *Art and Science in Breeding*, 186.

212 *"Modern science . . . threatens"*: Ibid., 177.

212 *When a bill to tighten inspections*: Riffel, *The Feathered Kingdom*, 116.

212 *By 1960, 95 percent of Arkansas*: Ibid., 121.

212 *While they muttered about*: Kendall M. Thu and E. Paul Durrenberger, *Pigs, Profits, and Rural Communities* (Albany: State University of New York Press, 1998), 150.

212 *Attempts by growers*: Ben F. Johnson, *Arkansas in Modern America, 1930–1999* (Fayetteville: University of Arkansas Press, 2000), 192.

212 *For the first time in American history*: "Poultry Production," U.S. Environmental Protection Agency, accessed March 22, 2014, http://www.epa.gov/oecaagct/ag101/ printpoultry.html.

213 *"It soon became obvious"*: Sawyer, *The Agribusiness Poultry Industry*, 52.

213 *Tyson's $10 million in net sales*: Riffel, *The Feathered Kingdom*, 147.

213 *But each bird weighed*: Gerald Havenstein, "Performance Changes in Poultry and Livestock Following 50 Years of Genetic Selection," *Lohmann Information* 41 (December 2006): 30.

213 *Hungry markets opened up overseas*: Poultry and Egg Situation, no. 205–222 (1960): 14.

213 *Upstart businessmen like Frank Perdue*: Richard L. Daft and Ann Armstrong, *Organization Theory and Design* (Toronto: Nelson Education, 2009), 44.

213 *"It takes a tough"*: Philip Scranton and Susan R. Schrepfer, *Industrializing Organisms: Introducing Evolutionary History* (New York: Routledge, 2004), 226.

213 *More than half of the nation's*: "Injustice on Our Plates: Immigrant Women in the

U.S. Food Industry," Southern Poverty Law Center, November 2010, accessed May 14, 2014, http://www.splcenter.org/get-informed/publications/injustice-on -our-plates.

213 *It is often ugly, low-paid*: "The Cruelest Cuts," *Charlotte Observer*, February 10–15, 2008, http://www.charlotteobserver.com/poultry/.

214 *Fifty years after the Chicken of Tomorrow*: Jeanine Bentley, "U.S. Per Capita Availabil-ity of Chicken Surpasses That of Beef," U.S. Department of Agriculture Economic Research Service, September 20, 2012, accessed March 22, 2014, http://www.ers .usda.gov/amber-waves/2012-september/us-consumption-of-chicken.aspx#.Uy3B jV76Tvo.

214 *By 2001, the average American*: *Profiling Food Consumption in America* (Washington, D.C.: U.S. Department of Agriculture, 2003), ch. 2.

214 *In 2012, Tyson recorded*: Tyson Foods, Inc., RBC Capital Markets Consumer & Retail Investor Day Presentation, December 6, 2012.

214 *The sign commemorates*: "Chicken of Tomorrow," University of Arkansas, accessed March 22, 2014, http://www.uark.edu/rd_vcad/urel/info/campus_map/454.php.

214 *The marker near Razorback Stadium*: "The John W. Tyson Building," Poultry Science, Dale Bumpers College of Agricultural, Food & Life Sciences, University of Arkansas, accessed March 22, 2014, http://poultryscience.uark.edu/4534.php.

11. Gallus Archipelago

215 *This last is touted*: "Cobb500," Cobb-Vantress, accessed March 22, 2014, http://www .cobb-vantress.com/products/cobb500.

215 *The Cobb 700 made*: "Cobb 700 Broiler Sets New Standard for High Yield," The Poul-try Site, October 1, 2007, accessed March 22, 2014, http://www.thepoultrysite.com /poultrynews/12958/cobb-700-broiler-sets-new-standard-for-high-yield.

216 *In 2010, as the backyard*: "CobbSasso," Cobb-Vantress, accessed March 22, 2014, http://www.cobb-vantress.com/products/cobbsasso.

216 *Three major breeding companies control*: D. L. Pollock, *View from the Poultry Breeding Industry* (Princess Anne, MD: Heritage Breeders, 2006).

216 *Based in Arkansas like its parent*: Cobb-Vantress homepage, accessed March 22, 2014, http://www.cobb-vantress.com/.

216 *More than three hundred U.S. breeder*: "Industry Economic Data, Consumption, Ex-ports, Processing, Production," U.S. Poultry & Egg Association, accessed March 20, 2014, http://www.uspoultry.org/economic_data/.

216 *In 1950, before the Chicken*: Ewell Paul Roy, "Effective Competition and Changing Patterns in Marketing Broiler Chickens," *Journal of Farm Economics* 48, no. 3, part 2 (August 1, 1966): 188–201, accessed March 22, 2014, http://www.jstor.org/stable /10.2307/1236327?ref=search-gateway:822036029853343f575f5e09154af4c5.

216 *In 2010, only forty-seven days were*: *The Business of Broilers: Hidden Costs of Putting a Chicken on Every Grill* (Washington, D.C.: Pew Charitable Trusts, 2013), http://www .pewenvironment.org/news-room/reports/the-business-of-broilers-hidden-costs -of-putting-a-chicken-on-every-grill-85899528152.

216 *New vaccines based*: "Flip-Over Disease," Poultry News, in cooperation with Merck, accessed March 22, 2014, http://www.poultrynews.com/New/Diseases/Merks/202600 .htm.

216 *During the 1990s*: Temple Grandin and Catherine Johnson, *Animals in Translation: Using the Mysteries of Autism to Decode Animal Behavior* (New York: Scribner, 2005), 69.

217 *An Israeli team recently*: "Bald Chicken 'Needs No Plucking,'" *BBC News*, May 21, 2002, accessed March 22, 2014, http://news.bbc.co.uk/2/hi/science/nature/2000003.stm.

217 *Academic researchers such as*: Ian J. H. Duncan and Penny Hawkins, *The Welfare of Domestic Fowl and Other Captive Birds* (Dordrecht, Netherlands: Springer, 2010).

217 *As a longtime meat eater*: Karen Davis, *Prisoned Chickens, Poisoned Eggs: An Inside Look at the Modern Poultry Industry* (Summertown, TN: Book Publishing Company, 1996); P. C. Doherty, *Their Fate Is Our Fate: How Birds Foretell Threats to Our Health and Our World* (New York: The Experiment LLC, 2012); Steve Striffler, *Chicken: The Dangerous Transformation of America's Favorite Food* (New Haven: Yale University Press, 2005).

218 *Founded in 1954 by broiler*: "History of the National Chicken Council," National Chicken Council, accessed March 22, 2014, http://www.nationalchickencouncil.org /about-ncc/history/.

218 *Senator Chris Coons of Delaware*: "Delaware Senator Chris Coons Announces Formation of US Senate Chicken Caucus; Georgia Senator Johnny Isakson to Co-chair," National Chicken Council, October 4, 2013, accessed March 22, 2014, http://www .nationalchickencouncil.org/delaware-senator-chris-coons-announces-formation -us-senate-chicken-caucus-georgia-senator-johnny-isakson-co-chair/.

219 *He arrived in 1974*: Bill Roenigk, interview by Andrew Lawler, 2013.

220 *Not long after my visit*: Stephanie Strom, "F.D.A. Bans Three Arsenic Drugs Used in Poultry and Pig Feeds," *New York Times*, October 1, 2013, accessed March 22, 2014, http://www.nytimes.com/2013/10/02/business/fda-bans-three-arsenic-drugs-used -in-poultry-and-pig-feeds.html?_r=0.

220 *Every week, they process*: Ann E. Byrnes and Richard A. K. Dorbin, *Saving the Bay: People Working for the Future of the Chesapeake* (Baltimore: Johns Hopkins University Press, 2001), 142.

221 *Chickens now produce 100 million*: Anna Vladimirovna Belova et al., "World Chicken Meat Market—Its Development and Current Status," *Acta Universitatis Agriculturae Et Silviculturae Mendelianae Brunensis* 60, no. 4 (2012): 15–30, doi:10.11118/ actaun201260040015.

221 *Steele's success, after all*: Matt T. Rosenberg, "Largest Cities Through History," accessed March 22, 2014, http://geography.about.com/library/weekly/aa011201a.htm.

221 *Saudi Arabian leaders*: "USDA International Egg and Poultry Review," U.S. Department of Agriculture, November 20, 2012, 15: 47.

222 *To feed this demand*: Carolina Rodriguez Gutiérrez, "South America Eyes an Optimistic Future," *World Poultry*, October 25, 2010, accessed March 22, 2014, http:// www.worldpoultry.net/Home/General/2010/10/South-America-eyes-an-optimistic -future-WP008070W/.

222 *Many American plants*: Delmarva Agriculture Data: Weekly Broiler Chicks Report, U.S. Department of Agriculture, March 6, 2011, 15: 11.

222 *The expanded Fakieh Poultry Farms*: "Fakieh Poultry Farms," Fakieh Group, accessed May 14, 2014, http://www.bayt.com/en/company/fakieh-group-1412762/.

222 *"The attitude of a gentleman"*: D. C. Lau, trans., *Mencius* (Harmondsworth, U.K.: Penguin, 1970), 53.

222 *In his* Remembrance of Things: Marcel Proust, *Swann's Way (Remembrance of Things Past, vol. 1)*, trans. C. K. Scott Moncrieff (n.p.: Digireads, 2009).

223 *He's agreed to with only*: Bill Brown, interview by Andrew Lawler, 2013.

224 *The 600 million chickens*: Peter Singer and Jim Mason, *The Way We Eat: Why Our Food Choices Matter* (Emmaus, PA: Rodale, 2006), 24.

225 *Continuing south, I pass*: "History of Temperanceville," The Countryside Transformed: The Railroad and the Eastern Shore of Virginia, 1870–1935, accessed March 22, 2014, http://eshore.vcdh.virginia.edu/node/1935.

225 *This is part of a Tyson plant*: "Firefighters Contain Weekend Fire at Tyson Plant," *Meat & Poultry*, December 2, 2013, http://www.meatpoultry.com/articles/news _home/Business/2013/12/Firefighters_contain_weekend_f.aspx?ID=%7B 9CA8E3E6-0DC7-4653-B5CF-B6ECB7578A2F%7D&cck=1.

225 *Instead of encountering*: Karen Davis, interview by Andrew Lawler, 2013.

226 *"The industry has created chickens"*: Temple Grandin and Catherine Johnson, *Animals Make Us Human: Creating the Best Life for Animals* (Boston: Houghton Mifflin Harcourt, 2009), 219.

226 *The litany of horrors*: "Chickens," United Poultry Concerns, accessed March 22, 2014, http://www.upc-online.org/chickens/chickensbro.html.

228 *It is a haunting refrain*: Peter Singer, *Animal Liberation: A New Ethics for Our Treatment of Animals* (New York: New York Review, 1975).

228 *Her words echo back*: Kerry S. Walters and Lisa Portmess, *Ethical Vegetarianism: From Pythagoras to Peter Singer* (Albany: State University of New York Press, 1999), 11.

228 *"No, for the sake of a little flesh"*: Ibid., 28.

228 *More recently, the writer J. M. Coetzee*: J. M. Coetzee and Amy Gutmann, *The Lives of Animals* (Princeton, NJ: Princeton University Press, 1999), 21.

229 *When a Western aid organization*: Kevin McDonald, interview by Andrew Lawler, 2012.

229 *Though linguists say this*: Vivienne J. Walters, *The Cult of Mithras in the Roman Provinces of Gaul* (Leiden, Netherlands: E.J. Brill, 1974), 119.

229 *With the exception of a brief period*: Steven Englund, *Napoleon: A Political Life* (New York: Scribner, 2004), 240.

229 *During the 1789 revolution*: "Living in the Languedoc: Central Government: French National Symbols: The Cockerel (Rooster)," accessed March 22, 2014, http://www .languedoc-france.info/06141212_cockerel.htm; Nicholas Atkin and Frank Tallett, *The Right in France: From Revolution to Le Pen* (London: I.B. Tauris, 2003), 43.

229 *France's official seal is*: "Living in the Languedoc."

229 *The bird appears on coins*: Lawrence D. Kritzman et al., *Realms of Memory: The Construction of the French Past* (New York: Columbia University Press, 1998), 424.

229 *One of the perks of that high office*: Pascal Chanel, interview by Andrew Lawler, 2013.

230 *"We wired to the Picassos"*: Alice B. Toklas, *The Alice B. Toklas Cookbook* (New York: Perennial, 2010), 92.

230 *That chicken is mentioned in*: Jon Henley, "Top of the Pecking Order," *Guardian*, January 10, 2008, accessed March 22, 2014, http://www.theguardian.com/environment/2008 /jan/10/ethicalliving.animalwelfare.

230 *The celebrated gastronomical*: Ibid.

230 *In 1862, a local count organized*: George W. Johnson and Robert Hogg, eds., *The Journal of Horticulture, Cottage Gardener, and Country Gentleman* (London: Google eBook, 1863).

230 *But a French colleague explained to*: W. B. Tegetmeier and Harrison Weir, *The Poultry Book: Comprising the Breading and Management of Profitable and Ornamental Poultry, Their Qualities and Characteristics; to Which Is Added "The Standard of Excellence in Exhibition Birds," Authorized by the Poultry Club* (London: George Routledge and Sons, 1867), 257.

231 *Every December, four contests*: "Les Glorieuses De Bresse, Votre Marché Aux Volailles Fines," accessed May 14, 2014, http://www.glorieusesdebresse.com/.

231 *Granted an* appellation d'origine contrôlée: Squier, *Poultry Science*, 159.

233 *About 250 farms raise 1.2 million Bresse*: "Producers' Portraits: Christophe Vuillot," Rungis, accessed March 22, 2014, http://www.rungismarket.com/en/vert/portraits_ producteurs/vuillot.asp.

233 *"The problem is to find"*: Georges Blanc, interview by Andrew Lawler, 2013.

234 *Though dressed simply in his chef whites*: Georges Blanc Vonnas Hotel Restaurant, Official Website, accessed March 22, 2014, http://www.georgesblanc.com/uk/index .php.

235 *Joyce was raised on*: Ron Joyce, interview by Andrew Lawler, 2013.

237 *She says that improving the lives of*: Leah Garces, telephone interview by Andrew Lawler, 2013.

12. The Intuitive Physicist

239 *In 1847, as The Fancy*: Barry Popik, " 'Why Did the Chicken Cross the Road?' (Joke)," *The Big Apple* (blog), August 28, 2009, accessed March 24, 2014, http://www .barrypopik.com/index.php/new_york_city/entry/why_did_the_chicken_cross_the _road_joke.

239 *In L. Frank Baum's 1907*: L. Frank Baum, *Ozma of Oz* (Ann Arbor: Ann Arbor Media Group, LLC, 2003).

239 *"Get away, yuh dumb"*: Edmund Wilson, *I Thought of Daisy* (New York: Farrar, Straus and Young, with Ballantine Books, 1953), 126.

239 *The term* birdbrain: *Online Etymology Dictionary*, s.v. "birdbrain," accessed March 22, 2014, http://www.etymonline.com/index.php?term=bird-brain.

239 *and* chickenshit were: *Urban Dictionary*, s.v. "chickenshit," accessed March 22, 2014, http://www.urbandictionary.com/define.php?term=chickenshit.

240 *The only thing dumber than*: William Lindsay Gresham, *Nightmare Alley* (New York: New York Review Books, 2010).

240 *The Italian neuroscientist Giorgio Vallortigara*: Giorgio Vallortigara, interview by Andrew Lawler, 2013. For his research papers, see "Research Outputs Giorgio Vallortigara," University of Trento, Italy, accessed March 22, 2014, http://www4.unitn.it /Ugcvp/en/Web/ProdottiAutore/PER0033020.

240 *Descartes declared that*: Lilli Alanen, *Descartes's Concept of Mind* (Cambridge, MA: Harvard University Press, 2003), 86.

241 *Chickens also use each eye*: Giorgio Vallortigara, "How Birds Use Their Eyes: Opposite Left-Right Specialization for the Lateral and Frontal Visual Hemifield in the Domestic Chick," *Current Biology* (January 9, 2001), 23.

241 *Recently, a research team*: Silke S. Steiger, "Avian Olfactory Receptor Gene Repertoires: Evidence for a Well-Developed Sense of Smell in Birds?" *Proceedings: Biological Sciences* 275, no. 1649 (October 22, 2008): 2309–317, accessed March 22, 2014, http:// www.jstor.org/stable/10.2307/25249806?ref=search-gateway:7c9847c5706d0eb705f 2b66ea9b88456.

241 *The bird also recalls faces of humans and chickens*: Carolynn L. Smith and Sarah L. Zielinksi, "The Startling Intelligence of the Common Chicken," *Scientific American* 310, no. 2, accessed March 22, 2014, http://www.scientificamerican.com/article/the -startling-intelligence-of-the-common-chicken/.

241 *"Even if chickens had a grammar"*: George Page, "Speak, Monkey," *New York Times*,

March 11, 2000, accessed March 22, 2014, http://www.nytimes.com/2000/03/12/books/speak-monkey.html.

241 *Since then, a German linguist*: Gail Damerow, *Storey's Guide to Raising Chickens: Care, Feeding, Facilities* (North Adams, MA: Storey Pub., 2010), 30.

242 *Chicks also are capable of adding*: R. L. Fontanari Rugani et al., "Arithmetic in Newborn Chicks," *Proceedings of the Royal Society B: Biological Sciences* 276, no. 1666 (2009): 2451–460, doi:10.1098/rspb.2009.0044.

243 *In 2012, an Australian*: Andy Lamey, "Primitive Self-Consciousness and Avian Cognition," ed. Sherwood J. B. Sugden, *Monist* 95, no. 3 (2012): 486–510, doi:10.5840/monist201295325.

244 *He founded the Hi-Bred*: Margaret Derry, *Art and Science in Breeding: Creating Better Chickens* (Toronto: University of Toronto Press, 2012), 143.

244 *American hatcheries churn out 500*: Zoe Martin, "Iowa Leads Nation in Many Ag Production Sectors," *Iowa Farmer Today*, March 13, 2014, accessed March 23, 2014, http://www.iowafarmertoday.com/news/crop/iowa-leads-nation-in-many-ag-production-sectors/article_63e4a5d6-aa01-11e3-9e9d-001a4bcf887a.html.

244 *The hens lay more than*: J. A. Mench et al., "Sustainability of Egg Production in the United States—The Policy and Market Context," *Poultry Science* 90, no. 1 (2010): 229–40, doi:10.3382/ps.2010-00844.

244 *Nine out of ten laying hens*: "Birds on Factory Farms," ASPCA, accessed March 23, 2014, http://www.aspca.org/fight-cruelty/farm-animal-cruelty/birds-factory-farms.

244 *Eight birds may be assigned*: Wilson G. Pond and Alan W. Bell, *Encyclopedia of Animal Science* (New York: Marcel Dekker, 2005), s.v. "Layer Housing: enriched cages."

244 *Vicious pecking, avian hysteria*: Donald D. Bell and William D. Weaver Jr., eds., *Commercial Chicken Meat and Egg Production* (New York: Springer, 2013), 92.

244 *Birds are debeaked*: Tom L. Beauchamp and R. G. Frey, *The Oxford Handbook of Animal Ethics* (Oxford: Oxford University Press, 2011), 756.

244 *Disturbing conditions like fatty*: Nuhad J. Daghir, *Poultry Production in Hot Climates* (Wallingford, CT: CABI, 2008), 202.

244 *"In no way can these living conditions"*: "Factory Egg Farming Is Bad for the Hens," Environmental Organizers' Network, accessed March 23, 2014, http://www.wesleyan.edu/wsa/warn/eon/batteryfarming/hens.html.

244 *"A gallinaceous madhouse"*: Roy Bedichek, *Adventures with a Texas Naturalist* (Austin: University of Texas Press, 1961), 115.

245 *Chickens are excluded*: Mench, "Sustainability of Egg Production in the United States."

245 *Banned in the European Union*: James Andrews, "European Union Bans Battery Cages for Egg-Laying Hens," *Food Safety News*, January 19, 2012, accessed March 23, 2014, http://www.foodsafetynews.com/2012/01/european-union-bans-battery-cages-for-egg-laying-hens/#.Uy7Hyl76Tvo.

245 *Costco and Walmart sell only*: "Barren, Cramped Cages: Life for America's Egg-Laying Hens," The Humane Society of the United States, April 19, 2012, accessed March 23, 2014, http://www.humanesociety.org/issues/confinement_farm/facts/battery_cages.html.

245 *A new $1.8 million research facility*: "Laying Hen Facility Opens New Doors for Research Achievements at MSU," ANR Communications, December 12, 2012, accessed March 23, 2014, http://www.anrcom.msu.edu/anrcom/news/item/laying_hen_facility_opens_new_doors_for_research_achievements_at_msu.

245 *Animal welfare scientist Janice*: Janice Siegford, interview by Andrew Lawler, 2013.
248 *Sheila Ommeh, who is tall*: Sheila Ommeh, interview by Andrew Lawler, 2013.

13. A Last Cause

251 *Chung is the protector of the*: Nguyen Dong Chung, interview by Andrew Lawler, 2013.
252 *That fact worries the Chinese biologist*: Han Jianlin, interview by Andrew Lawler, 2013.
253 *One-third of South Asians*: "Highest Prevalence of Malnutrition in South Asia," *Bread for the World*, June 25, 2012, accessed March 23, 2014, http://www.bread.org/media /coverage/news/highest-prevalence-of.html.
253 *Chicken meat and eggs*: David Farrell, "The Role of Poultry in Human Nutrition," *Poultry Development Review*, Food and Agricultural Organization of the United Nations, 2013, http://www.fao.org/docrep/019/i3531e/i3531e.pdf.
254 *The country's rapid development*: "Vietnam Population 2014," World Population Review, April 2, 2014, accessed March 23, 2014, http://worldpopulationreview.com /countries/vietnam-population/.
254 *The number of birds*: Le Thi Thuy, interview by Andrew Lawler, 2013.
254 *After China, Saudi Arabia, and Singapore*: "Global Poultry Trends 2013: Asia Puts More Emphasis on Trading Prepared Products," The Poultry Site, September 11, 2013, accessed March 24, 2014, http://www.thepoultrysite.com/articles/2894/global -poultry-trends-2013-asia-puts-more-emphasis-on-trading-prepared-products.
255 *Recent research suggests*: T. M. Tumpey, "Characterization of the Reconstructed 1918 Spanish Influenza Pandemic Virus," *Science* 310, no. 5745 (2005): 77–80, doi:10.1126 /science.1119392.
255 *In the late 1990s*: K. M. Sturm-Ramirez et al., "Reemerging H5N1 Influenza Viruses in Hong Kong in 2002 Are Highly Pathogenic to Ducks," *Journal of Virology* 78, no. 9 (2004): 4892–901, doi:10.1128/JVI.78.9.4892–4901.2004.
255 *All 1.6 million of the city-state's*: Paul K. S. Chan, "Outbreak of Avian Influenza A(H5N1) Virus Infection in Hong Kong in 1997," *Clinical Infectious Diseases* 34, no. Supplement 2: Emerging Infections in Asia (May 01, 2002), accessed March 23, 2014, http://www.jstor.org/stable/10.2307/4483086?ref=search-gateway:d5ff295abb 4b1cd99e8ddb8a5ee36030.
255 *"From the tips of their combs"*: John Farndon, *Bird Flu: Everything You Need to Know* (Thriplow, U.K.: Icon, 2005), 9.
255 *In 2004, Thai prime minister*: Chanida Chanyapate and Isabelle Delforge, "The Politics of Bird Flu in Thailand," Focus on the Global South, accessed March 23, 2014, http://focusweb.org/node/286.
255 *In 2013, a new strain of the H7N9*: Robert J. Jackson, *Global Politics in the 21st Century* (Cambridge: Cambridge University Press, 2013), 506.
256 *Some researchers argue that*: C. Larson, "Tense Vigil in China as Nasty Flu Virus Stirs Back to Life," *Science* 342, no. 6162 (2013): 1031, doi:10.1126/science.342.6162.1031.
256 *This radical approach*: Hongjie Yu et al., "Effect of Closure of Live Poultry Markets on Poultry-to-Person Transmission of Avian Influenza A H7N9 Virus: An Ecological Study," *Lancet* 383, no. 541 (February 8, 2014).
256 *Other researchers say that disinfecting*: Q. Liao and R. Fielding, "Flu Threat Spurs Culture Change," *Science* 343, no. 6169 (2014): 368, doi:10.1126/science.343.6169.368-a.
256 *"This feature of chicken"*: Page Smith and Charles Daniel, *The Chicken Book* (Boston: Little, Brown, 1975), 194.

257 *Only farmed salmon*: "Salmon Have the Most Efficient Feed Conversion Ratio (FCR) of All Farmed Livestock," *Mainstream Canada*, accessed March 24, 2014, http://msc.khamiahosting.com/salmon-have-most-efficient-feed-conversion-ratio-fcr-all-farmed-livestock.

257 *Meat production accounts*: Henning Steinfeld, *Livestock's Long Shadow: Environmental Issues and Options* (Rome: Food and Agriculture Organization of the United Nations, 2006), 112.

257 *The British economist Thomas Malthus*: Thomas Robert Malthus, *An Essay on the Principle of Population* (Cambridge: Cambridge University Press, 1989).

257 *A billion chickens*: "Global Poultry Trends 2013: Chicken Imports Rise to Africa, Stable in Oceania," The Poultry Site, November 13, 2013, accessed March 23, 2014, http://www.thepoultrysite.com/articles/2972/global-poultry-trends-2013-chicken-imports-rise-to-africa-stable-in-oceania.

257 *Global poultry exports*: Ibid.

257 *U.S. consumers eat four times*: "Per Capita Consumption of Poultry and Livestock, 1965 to Estimated 2014, in Pounds," National Chicken Council, last modified January 10, 2014, accessed March 23, 2014, http://www.nationalchickencouncil.org/about-the-industry/statistics/per-capita-consumption-of-poultry-and-livestock-1965-to-estimated-2012-in-pounds/.

257 *Mexicans are the leading egg eaters*: Carrie Kahn, "It's No Yolk: Mexicans Cope with Egg Shortage, Price Spikes," NPR, September 18, 2012, accessed March 23, 2014, http://www.npr.org/blogs/thesalt/2012/09/18/160968082/its-no-yolk-mexicans-cope-with-egg-shortage-price-spikes.

257 *In China, people eat twenty-two pounds*: Janet Larsen, "Plan B Updates: Meat Consumption in China Now Double That in the United States," April 24, 2012, accessed March 23, 2014, http://www.earth-policy.org/plan_b_updates/2012/update102.

258 *In 2012, for the first time*: Ibid.

258 *Fujian Sunner Development Co.*: "Fujian Sunner Development Co., Ltd. Class A," Morningstar, accessed March 23, 2014, http://quicktake.morningstar.com/stocknet/secdocuments.aspx?symbol=002299&country=chn; "China Rich List: #151 Fu Guangming & Family," *Forbes*, accessed March 23, 2014, http://www.forbes.com/profile/fu-guangming/.

258 *Tyson Foods intends to build ninety*: David Kesmodel and Laurie Burkitt, "Inside China's Supersanitary Chicken Farms," *Wall Street Journal*, December 9, 2013, accessed March 23, 2014, http://online.wsj.com/news/articles/SB10001424052702303559504579197662165181956.

258 *In 2014, the Chinese*: Ian Johnson, "China Releases Plan to Incorporate Farmers into Cities," *New York Times*, March 17, 2014, accessed March 23, 2014, http://www.nytimes.com/2014/03/18/world/asia/china-releases-plan-to-integrate-farmers-in-cities.html?_r=0.

258 *By heating, cooling, and pressurizing*: Alton Brown, "Tastes Like Chicken," *Wired*, September 15, 2013, accessed March 23, 2014, http://www.wired.com/wiredscience/2013/09/fakemeat/.

258 *Another California startup*: "Hampton Creek, Named by Bill Gates as One of Three Companies Shaping the Future of Food, Debuts First Product at Whole Foods Market," Business Wire, September 20, 2013, accessed March 23, 2014, http://www.businesswire.com/news/home/20130920005149/en/Hampton-Creek-Named-Bill-Gates-Companies-Shaping.

259 *The chicken in China*: Fang Zhang, *Animal Symbolism of the Chinese Zodiac* (Beijing: Foreign Languages Press, 1999), 100.

260 *"Don't try to convey your enthusiasm"*: Roy Edwin Jones, introduction in *A Basic Chicken Guide for the Small Flock Owner* (New York: W. Morrow, 1944).

260 *Upscale Neiman Marcus*: "Beau Coop," Neiman Marcus, accessed March 23, 2014, http://www.neimanmarcus.com/christmasbook/fantasy.jsp?cid=CBF12_O5415& cidShots=m,a,b,c,z&r=cat44770736&rdesc=The%20Fantasy%20Gifts&pa geName=Beau%20Coop&icid=CBF12_O5415.

260 *Just as the Victorian Fancy cooled*: Jonel Aleccia, "Backyard Chickens Dumped at Shelters When Hipsters Can't Cope, Critics Say," *NBC News*, July 7, 2013, accessed March 23, 2014, http://www.nbcnews.com/health/health-news/backyard-chickens -dumped-shelters-when-hipsters-cant-cope-critics-say-f6C10533508.

261 *"The snare is carried in a basket-like case"*: Fay-Cooper Cole and Albert Gale, *The Tinguian: Social, Religious, and Economic Life of a Philippine Tribe* (Chicago: Field Museum of Natural History, 1922).

262 *Nguyen Quir Tuan*: Nguyen Quir Tuan, interview by Andrew Lawler, 2013.

264 *Brisbin, the man who saved Gardiner*: I. Lehr Brisbin, interview by Andrew Lawler, 2013.

BIBLIOGRAPHY

Please see www.andrewlawler.com for a complete list of references.
http://bit.ly/WhyDidtheChickenCrosstheWorld_Biblio

INDEX

ABOUT THE AUTHOR

Andrew Lawler landed his first job as a reporter just days before the 1986 *Challenger* space shuttle disaster and spent the next decade and a half covering science and technology politics as a staff writer for several publications in Washington. After a year at MIT as a Knight Science Journalism fellow, he founded the New England bureau of *Science* and began reporting frequently on archaeology in the Middle East, Central Asia, and the Far East. His articles and op-ed pieces have appeared in more than a dozen publications, including *National Geographic*, *Smithsonian*, *The New York Times*, *Audubon*, *Slate*, and *The Washington Post*, as well as in *The Best of American Science and Nature Writing*. Lawler is a contributing writer for *Science* and a contributing editor for *Archaeology*. He owns no chickens, preferring to visit those of his friends. For more, see www.andrewlawler.com.